FLY THE WILD AND STAY ALIVE

A Bush Flight Training Manual

Hal Terry

Best of luck!
Hal Terry

WORDPRO PRESS

Published in 2000 by TERRY AVIATION ENTERPRISES.

0-9673116-6-7

TABLE OF CONTENTS

CHAPTER SEVEN – SECURING YOUR AIRPLANE; TAKING OFF

CHAPTER EIGHT – WILDERNESS SURVEY PILOT PROCEDURES AND TECHNIQUES

CHAPTER NINE -- SURVIVAL FLYING IN EXTREME CONDITIONS

FOREWORD

A basic premise of this book is that the Private Pilot, by hard practice and continuing evaluation (by self and others), can improve his or her piloting skills to the point that bush flying risks will be minimized. The author bases this belief on his own experiences in bush flying and in military flying.

Just before the Korean War, the author was twenty-two and a U.S. Navy pilot in a Pacific Fleet fighter squadron. Flying the Grumman F8F-2 "Bearcat," his outfit was Fighter Squadron 191, known as "Satan's Kittens." Within a few days of its return from an extended carrier deployment to the western Pacific, war erupted in Korea.

Within days, the squadron started a rapid transition to the Grumman F9F-2 "Pantherjet." Simultaneously, the Navy's entire flight demonstration team, the Blue Angels, was assigned to the squadron. Lieutenant Commander Johnny Magda, Leader of the Blues, became skipper of Satan's Kittens. What a great skipper he was, and what a great bunch of men (both pilots and supporting crew) the Blues were. Young and impressionable, the author tried to pattern himself after them, both in demeanor and in flying airplanes.

To a man, the ex-Blues pilots demonstrated two unspoken rules. First: only hard, continuing practice makes for near-perfect performance. And second: each pilot must be self-critical of his own

performance and ready to accept and act on criticisms coming from others. They seemed to accept themselves as average pilots continuing to learn instead of pilots who had "arrived." Obviously, they believed there's always room for anyone to improve.

The author incorporated those rules in his flying and pursued the "trying hard to get better" concept from then on. Fifteen years later, the Navy offered him the position of Blue Angels Leader. But his wife recently had died of a virus and he had to look after four small children and not be gone so much.

The two lessons learned from the Blues -- hard practice makes near-perfect and always keep trying to improve -- had enabled him to achieve the level to at least be offered the job he had wanted so much. In the same way, one can achieve the goal of being a skillful, safer bush pilot.

The Private Pilot, by practicing hard and always striving to improve, can achieve and maintain that goal. It also worked for the author during ten thousand flight hours in the Alaskan bush. It can for you, too.

Author Information

Harold L. (Hal) Terry was a Navy pilot for thirty years, flying mostly fighters and attack aircraft. He flew over sixty different models of Navy aircraft, from the Grumman F6F Hellcat and F8F Bearcat and the Chance Vought F4U Corsair to the Vought F-8 Crusader and McDonnell F-4 Phantom II. His shore duty included time as an advanced fighter training flight instructor and as a weapon systems test pilot. He flew in the Korea and Vietnam wars and was skipper of a fighter squadron and CAG of a carrier air wing in combat.

Shortly after retiring from the Navy he became a CFI, and is qualified to teach in land and sea airplanes, single-engine and multi-engine, and instruments. He has flown seventy different models of civilian airplanes. Hal holds an ATP license for airplanes, multi-engine land and sea, and single-engine land, with Commercial Pilot privileges for single-engine sea and the T-33B. He is an A&P Mechanic with Inspection Authorization. In 1980 he taught at the Ft. Richardson, Alaska, Flying Club, then flew air taxi floatplanes at Kodiak, Alaska and subsequently managed Denali Wilderness Air at McKinley Park. He then was a pilot/mechanic and regional check pilot/instructor for the Alaska Dept. of Fish and Game, Kodiak, and also an FAA-designated Pilot Examiner and Safety Counselor. He retired from ADF&G in 1993, and has since been Chief Pilot (Part 135), Chief CFI (Part 141), and Director of Maintenance (Part 135) for Leading Edge Aviation in Tucson. Hal has over 19,000 hours.

Acknowledgements

Early in my wilderness flying, I found no books on bush flying techniques. So, I began making notes. Friend and owner of Denali Flying Service (later Denali Wilderness Air), Master Guide and bush pilot Lynn Castle strongly supported my efforts. Later, he lost his life in a mountain bush flying accident. This book is dedicated to Lynn's memory.

Alaskan pilots who likely do not agree with <u>everything</u> herein, but who have given support of this effort include Larry Nicholson (my bush-flying ADF&G boss/now a CFI/air taxi bush pilot), and Ralph Wright (master bush pilot/chief pilot/chief mechanic of Flirite and ADF&G days). Thanks also to Jacque Bunting (Pilot Examiner/CFI/air taxi bush pilot), John Riley Morton (war hero, Safety Counselor, CFI, and air taxi bush pilot), Gene Storm (my editor, Air Alaska magazine), and Tom Wardleigh and Ginny Hyatt (leaders of the great Alaskan Aviation Safety Foundation). For his review and comments on the final manuscript, I also thank long-time Alaskan bush pilot, Master Guide, and former Governor of Alaska, Jay S. Hammond.

Thanks, certainly, to my wife, Bonnie, who endured with grace my eternal muttering while processing these words.

Finally, many thanks to WORDPRO Press (on the Internet) for making it possible for a book of interest to a relatively small market to be published.

CHAPTER ONE -- HONING THE REQUIRED SKILLS AND ATTITUDES

Basic Requirements of Safer, Professional-Level, Bush Piloting

Safer bush flying requires a high degree of flight control coordination and "feel" for the airplane in slow flight. It takes awareness of your situation such as navigation, weather, fuel, etc. It also takes a good eye for sizing-up prospective off-airport operation areas from the air. That is, you have to be good at flying slow, at realizing what's going on around you and what it means, and at airborne survey of prospective surface areas.

Prior to an off-airport landing you must have a plan for landing, re-positioning, and the subsequent takeoff and departure. And you need the stick-and-rudder skills to extract the maximum capabilities from your airplane. You don't have to be fancy, just plain good with the basics. Here's an important thing: you simply have to keep trying to do your best, no matter how experienced you become. That's because no matter how experienced you are, you always can be better. Along with stick-and-rudder skills, the ability to remain aware of your situation, and the ability to evaluate surface areas, you also need enough common sense to say "no" when it's appropriate.

For example, you have to be able to make that most difficult turn of all -- the 180-degree turn -- in the face of external pressures and your own internal urge to press on despite a deteriorating situation. It's a fact: bush flying presents the pilot with frequent opportunity for life-or-death decisions. Often a pilot doesn't realize, until later, how important a decision was to his continued good health. What made the difference was plain, dispassionate common sense. (Figure 1-1)

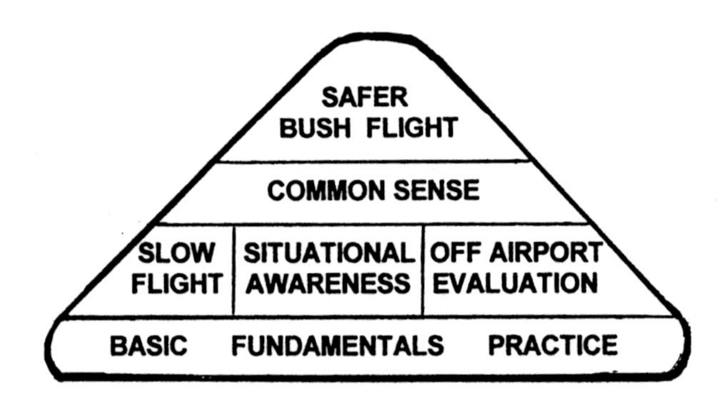

ACHIEVING SAFER BUSH FLIGHT

FIGURE 1-1

Safe bush flying requires higher skill levels than those required for a Private Pilot's license. The author did not fully realize this until he became a regional flight instructor/check pilot for the Alaska Dept. of Fish and Game. During the same period of time, as an FAA-designated pilot examiner, he gave practical tests when he was "in town" at Kodiak in the winter. Both pursuits were collateral to a year-round job as pilot/mechanic for Fish and Game in

the eastern Aleutians, Alaska Peninsula, and Kodiak Island areas.

Evaluating pilots against two different standards highlighted the difference in skill levels required for safe bush flying compared to those required for the Private Pilot certificate. As stated in the Foreword, with continuing work and by always trying to get better, the Private Pilot can achieve the higher levels required for safer bush flying. In fact, the bush flying skills emphasized herein will make <u>any</u> pilot a much better, much safer pilot—this, without ever going near the wilderness.

Four Kinds of Bush Pilots

This author believes that pilots who fly the bush, and those who are in the "wannabe" stage, come in four basic varieties.

First, there are the bold, brash, adrenaline junkies (amateurs and a few pros) who seem to be out to prove something. Sooner or later they will prove that old, <u>bold</u> bush pilots are scarce as hen's teeth.

Second are the pilots who seem to approach flying the way they drive their autos—sometimes skillfully, yet a tad too cavalier about small but important things.

Third are those amateurs who treat bush flying as a highly technical sport that takes constant effort to improve both knowledge and skills.

And fourth are most of those who make a living at flying the bush. Their frequent exposure

ADRENALINE JUNKIES	THE CAVALIER
SKILLED AMATEURS	REAL PROS
OTHER PILOTS INTERESTED IN BUSH FLYING	

THIS BOOK IS FOR ...

FIGURE 1-2

develops within them the right blend of skill, situational awareness, and disciplined mental attitudes. They're not super-human. They make mistakes like everyone else, but they are constructively critical of their own performance. That trait makes them continue to improve. For them, flight hours are more than a number; they represent cumulative lessons in wilderness flight.

There are, of course, many combinations of the above four pilot categories. This book is for all four, hoping that this can help the people in the first and second categories survive long enough to mature into solid third and fourth category bush pilots.

In flying, as in many skills, there are usually several different ways to "skin a cat." Therefore, those conscientious pilots of the third and fourth categories may find some different but useful ideas here.

There is a fifth, non-bush category of pilots who will find this book useful. They are those who always seek new knowledge and skills based on other pilots' experiences. These pilots will enjoy acquiring this knowledge and perhaps might even take steps to improve the skills emphasized herein. One never knows, some may even find themselves becoming a safe bush pilot of the third or fourth category. Some of the most beautiful and soul-searing flight can be found only in the wilderness. You must do it to believe it.

How Much Skill Is Required For Bush Operations?

At pilot meetings, where I've talked on bush flying, some have asked for numbers on my assertion that the bush pilot who wants to be safer must have particular pilot skills above those at the Private Pilot level. The numbers that follow have to do with how skillful a pilot ought to be in takeoffs and landings. (That's because a high percentage of aircraft accidents happen during these basic operations in the bush.)

FIRST, the safer bush pilot ought to be able to touch down within 50 feet longitudinally and three feet laterally of a chosen touchdown point. The plane ought to be lined up, with no lateral drift, in cross-wind components up to at least 20 percent of the wings level, flaps down stall speed. The pilot should do this consistently. (Figure 1-3 next page.)

Now, before the "silver backs" start howling: the author does not advocate landing where you need to touch down within 50 feet lengthwise and 3 feet sidewise or you'll bend your airplane. We're talking skill levels here - the skill levels you should have. But, in the real bush situation, give yourself as much extra room as you can. You <u>always</u> want a comfortable safety margin. But you need the higher skill levels to handle the unexpected things that happen frequently in the boon-docks.

THE AIM POINT AND INTENDED TOUCH DOWN

FIGURE 1-3

SECOND, in the event of a bounced landing, and with the above maximum crosswind, the pilot should be able to re-acquire the "center line" <u>and</u> re-establish drift correction -- all this with little or no addition of power. (On a one-way strip, by definition, you won't be able to go around.) Think about the sequence that takes place after your low (upwind) main wheel hits a bump and you bounce with a wing down into a crosswind. You'll agree that things can get real sporty by the time you regain control and resume non-drifting line-up over the landing area "center line." Doing that successfully on a one-way strip takes quick recognition of what's going on and what to do with the flight controls and throttle. (Figure 1-4.)

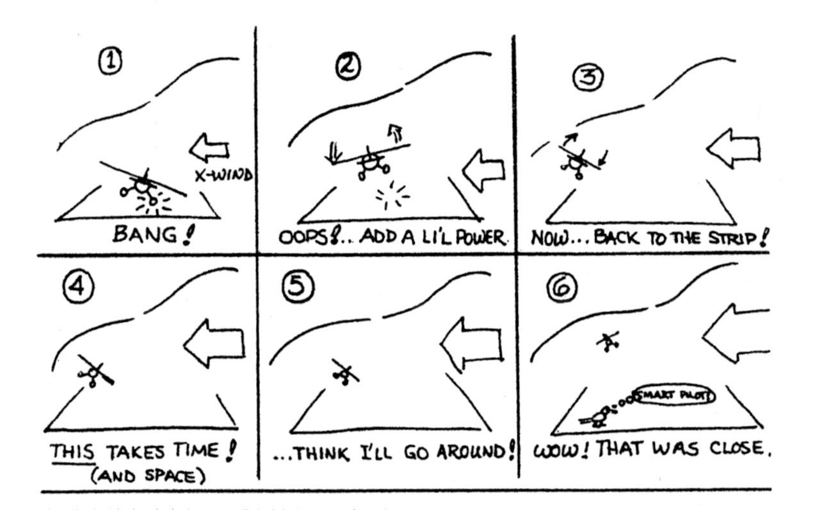

THE CROSS WIND WING-DOWN BOUNCE
FiGURE 1-4

THIRD, during any landing approach in fairly smooth air, be able to hold the airspeed within two knots/mph of pilot's handbook approach speed in a steady manner, down to the point where slowing and transition to the flare begins. Given a stabilized glide slope, this is attainable with prompt but small throttle corrections.

If there is any doubt about the accuracy of the aircraft's pitot-static system, you should have the system and airspeed indicator checked by a qualified shop.

Some small airplanes have large differences between IAS and CAS, especially at low airspeeds, due to "position" and instrument errors. Also, these errors vary a lot across the low airspeed range. Because of that, do not figure your approach speed based on indicated airspeed at stall. It should be based on calibrated airspeed. This is why it's important to fly handbook approach airspeeds. Some older models give operating speeds in CAS, thus requiring you to use the IAS/CAS tables to obtain useful IAS numbers.

Some models of aircraft [including some with short-take-off-and-landing (STOL) modifications] will demonstrate an impressive rate of descent <u>prior</u> to the stall break. There might <u>not</u> even be a stall break. What can happen here is that you "run out of elevator" before wing airflow separation.

On the other hand, elevator control could be okay but lift-killing airflow separation develops before the stall break and descent begins early. The net effect is the same, with power at idle you will not be able to touch down in <u>level</u>, controlled flight below

that indicated airspeed.

FOURTH in the skills department, be able to estimate your landing roll stopping point, takeoff lift-off point, and vertical obstacle clearance; the surface rolls within 200 feet and the vertical clearance within 10 feet. This takes plenty of practice. So, make a habit of doing it on every landing and takeoff from now on.

Good Performance Estimates Will Help Most To Keep You Out Of Trouble

Of all the above criteria, the fourth is a shade more important than the others. Being able to estimate accurately your performance with a particular airplane and undercarriage combination over a wide range of operating surfaces, winds, and load conditions, as much or more than anything else will minimize the risks *always* inherent in off-airport operations. Your ability to estimate will continue to improve if you make it a part of every takeoff and landing routine.

Serious Risks Are Always Present In Bush Flying

Certainly, there are serious risks in all facets of bush flying, no matter how skillful the pilot is. Accurate, conservative estimates will help a lot in your management of those risks. The value of common sense is often more evident later.

A Self-Training Program Assisted By The Flight Instructor

How can you achieve the required level of

expertise? Here's a six-point, assisted self-training program you can put into your routine flying and it won't take long to achieve your goals. <u>Important! Having an experienced CFI fly with you frequently in this program will make your training safer, more effective, and less expensive in the long run.</u>

FIRST, as much as possible, and as density altitude and the operating surfaces allow, fly your plane at its maximum allowable landing gross weight. The idea is to get used to the reduced (maybe even lousy) performance of a working airplane. By doing this you will learn the pitch attitude/power combinations that will give you maximum performance with precise V-speeds (Vx and Vy). Without this, you can't be ready to fly those loads of moose meat out of a small operating area. You need to have "been there, done that."

SECOND, if you're rusty on maximum performance takeoffs and climbs, and/or short/soft landings, take a few rides with a CFI to get your <u>procedures</u> "down pat." (See Appendix B, this chapter.) After developing your procedures, it's a matter of getting the accuracy, feel, smoothness, and consistency that comes from self-critical practice and experience. Make your predictions of airplane performance on <u>every</u> takeoff and landing. Predictions should become part of your flying. Constantly and critically evaluate yourself and you will improve. If you don't, you won't. Guaranteed.

THIRD, become an expert on the wheels landing and holding the tail wheel off the surface for a good portion of the roll-out. (More on this technique in a later chapter.) Learn how to do it without excess airspeed. Avoid the nose-down "strafing run wheely" in favor of touching down just a "hair" on the back side of the main wheels, with minimum sink rate and at your intended spot. This takes fine-tuned hands on the power and pitch controls -- not easy without practice.

FOURTH, get honed up on your stabilized approach. Being stabilized does <u>not</u> mean having a long, strung-out final. In fact, long, strung-out finals are hazardous in many situations. We're talking about arriving at the transition point (for power reduction and flaring to your touchdown) with all the variables steadied out. That is, you're on steady glide path, airspeed, pitch attitude, and lined up on "centerline" with no drift. You can think of it as being in a "safety cone" for arrival at the rather precise point of transition for touchdown. There will be acceptable (but less desirable) situations where the "cone" will be close-in, with a shallow banked turn. **(Figure 1-5, next page.)**

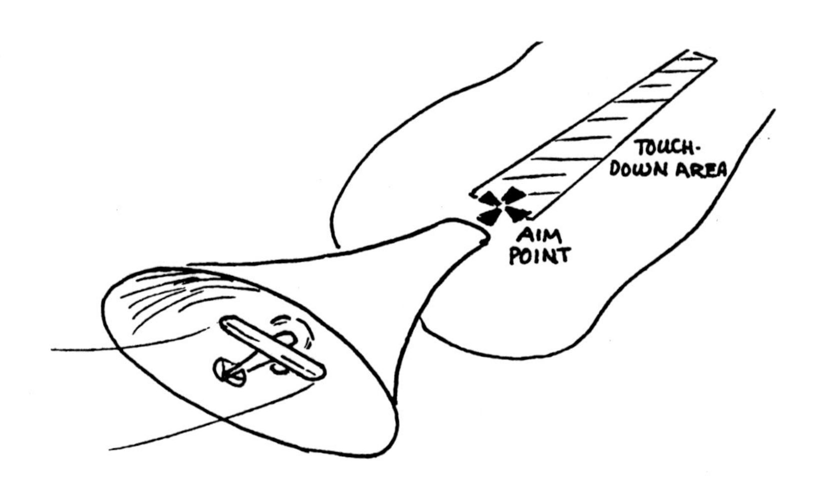

THE SAFETY CONE

FIGURE 1-5

Being stabilized makes your decisions for power reduction and flare much easier to take and more accurate. You're looking to stabilize your approach by making prompt (therefore small) and smooth pitch and power corrections. If the air is rough, you'll still be prompt but you won't be so smooth due to the size of upsetting gusts or shears and the required size of your corrections. (See the discussion in the following pages concerning landing approach flight control principles.)

FIFTH, have a specific surface aim-point on every landing approach, no matter how wide or long the landing area or runway is. The aim-point is the place where your visual glide slope intersects the surface. On a steady glide path, the aim-point will maintain a steady angular relationship with the nose

of your airplane down almost to the point of power reduction. If you didn't flare for touch down, the landing gear would touch down short of your aim-point. How short would depend on how far below your eye level the landing gear hangs and on your angle of descent. (Figure 1-6)

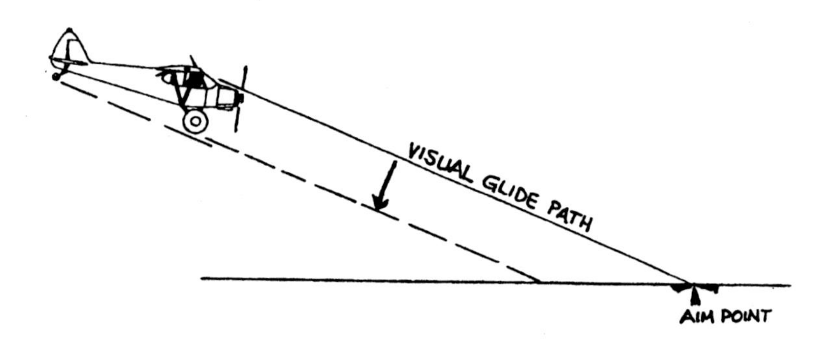

LANDING GEAR UNDER-HANG

FIGURE 1-6

You always want to have the aim-point up on the intended landing surface, to allow for the unexpected. If you're not too fast, this bit of safety factor will cost you only one or two plane lengths in roll-out. It's worth it.

SIXTH, always have at least one-and-a-half degrees or more of glide slope for your landing approaches in the off-airport environment. The steeper your glide path, the easier it is for you to detect a change in your projected touch down point. But your glide path must be shallow enough to require some power. Being able to carry some power allows you to coordinate pitch with power to maintain an on-speed glide path with precision.

The power-off glide, which depends on use of pitch with slips as necessary and/or landing flaps, is not conducive to the solid precision you're after. You'll do well to have two glide slope angles in your bag of skills; the no-obstacle one-and-a-half degree glide and the standard three degrees for use with approach obstacles. The high drag airplane, such as a float plane, permits the more desirable three degrees, power-on approach with precision.

The Drag-and-Drop Nonsense

Some pilots will prefer to have an aim-point well short of the landing area. Then, as they reach that point, with full flaps, they come on with power and "hang 'er on the prop," right up to where they want to land, and then chop the throttle. This is called the "drag-and-drop" method. Drag-and-drop is unnecessary and hard on airplanes. It is unsafe.

Think about it. Just when you need to see ahead and need stable airspeed for precision and safety, the drag-and-drop method has you raising the nose, blanking out your forward view and slowing down to put yourself at the mercy of wind shears and down drafts. That's not smart. (It's true that drag-and-drop is used in spot landing contests, and such contests do have merit. However, not too many of those contests are held in gusty cross winds at off-airport sites, with only the wild, hungry carnivores watching with interest.)

The argument that drag-and-drop is necessary for short areas doesn't make sense. That's because on unimproved surfaces, due to landing gear surface drag, you almost always need more room to get airborne than you need to land. Except for rough

water floatplane landings, drag-and-drop has no redeeming features. So, join the move to drop drag-and-drop.

A NOTE ON FLIGHT CONTROL PRINCIPLES FOR THE LANDING APPROACH

"Since short field approaches are power-on approaches, the pitch attitude is adjusted as necessary to establish and maintain the desired rate or angle of descent, and power is adjusted to maintain the desired airspeed. However, a coordinated *combination of both pitch and power adjustments* is usually required." [Italics <u>not</u> added.] FAA Flight Training Handbook, Revised 1980, AC 61-21A, p.111.

One argument easy to start among pilots is the one about "What controls airspeed and what controls altitude or angle of descent in a landing approach?" The author was lucky to have learned the best way, as described in the above-quoted words. It served me well in over twenty years as an active Navy carrier pilot.

Certainly, Navy Landing Signal Officers (LSO's) will preach "power" to correct for a low, especially in-close near the carrier's ramp. There is usually a down draft just aft of the carrier's ramp and the LSO wants to avoid being part of the fireball when any airplane looks ready to sink below the glide-slope. He therefore will rely on the fact that a sudden surge in engine thrust will cause a quick lift in the plane's glide path due to the engine's upward thrust line. But for most of the approach, a smooth coordination of pitch with power is required. Indeed, the smooth, coordinated approach still

works, be it aircraft carrier, ocean beach, or small lake. (On occasion, I have had to correct privately an LSO when he waxed eloquently and expanded the meaning of the close-in power call to that of pitch-power flight control principles in general.)

Aerodynamics tells us that most answers to the pitch/power question are over-simplified. Despite that, the question is very pertinent to flight training in the sense of <u>what CFI's ought to teach students</u>. In that sense, the FAA Flight Training Handbook says <u>exactly</u> what should be said for the power-on approach.

All you have to do is think of the most critical situations in the landing approach and then ask yourself what you would want a student pilot to do with the stick and/or throttle, and in what sequence, to effect a prompt yet safe recovery.

For example, suppose the fledgling finds himself flaring too early and too high and is twelve feet above the runway, nose high, just above stall speed? Does he pitch nose down to the proper airspeed or add power to accelerate? The result of pitching the nose down could be disaster. The best answer is, as stated in AC 61-21A, to use a coordinated combination of pitch and power, and the properly trained pilot will <u>lead</u> with power because the most immediate problem is airspeed.

That pilot will also make a smoothly coordinated pitch reduction so as to convert the added power to a speed increase rather than climb and allow the plane to remain critically slow. The pilot can then decide whether continued landing or go-around is possible and appropriate.

A solid understanding of pitch/power coordination as well as directional and lateral drift control is very important to the safety of any pilot but is especially important to the bush pilot.

To achieve a stabilized approach, almost always you should slow to approach speed after getting into landing configuration, acquire your aim-point, pitch to, and hold, the desired glide slope with elevator while making a coordinated power reduction. Then, you can fine-tune your airspeed with the power, while fine-tuning your elevator trim. In a turning approach with short final leg, you'll need slightly increased power, to be smoothly reduced as you roll out of the turn to final. Avoid more than 20-25 degrees of bank in that turn to final. If you're going to need more, go around early and get a better start.

As a safer bush pilot, you'll want to be alert for directional (heading) and lateral (sidewise) drift. You'll have to do that without the converging-line perspective of runways to help. Prior to the flare, switch your visual focus back and forth between the area just ahead and that farther down/up the landing area, to maintain good height reference. This is essential when landing on an uphill or downhill area. On a steeply rising runway, you will likely have to add power as you flare above the landing area. It's a "whatever it takes" situation. I have frequently had to go almost to climb power in a fully loaded Cessna U-206 when flaring uphill to a steep landing area a few miles south of Gold King Creek in interior Alaska. On those kinds of approaches, the pilot must realize the commitment to landing occurs

early in the approach. It can happen earlier than you expect.

During the final part of your flare, pick an area near the far end of the landing area to maintain directional alignment while using your peripheral vision to detect lateral crosswind drift during the transition to touchdown. Try to ensure that your crosswind technique is impeccable. You can do this only by having a CFI or highly experienced pilot ride along with you to help with analysis and critique.

In Appendix A to this chapter, the author includes a description of the "Precision Dutch Roll" training maneuver which will help you become skillful in the cross-control techniques required for wing-down crosswind landings.

Finally, using the training program and Appendices A and B as a guide, when you think you've got a handle on your techniques, go flying with an experienced bush pilot to confirm your achievement.

Before you launch into the wilds, it's worthwhile to take a look at your airplane and its basic equipment. We'll cover that somewhat in the next chapter, but first, some words about the foundation for decision-making in the pilot's seat: mental attitude.

The Five Hazardous Mental Attitudes

All the best procedures and techniques in the world cannot protect the pilot who lets hazardous mental attitudes influence decisions. That pilot will have to depend on luck. We all harbor, to varying

degrees, potentially hazardous mental attitudes. Otherwise, we wingless earthlings would not be trying to act like birds. Here are five serious, hazardous attitudes we must recognize and defeat. They are taken from an FAA safety pamphlet and listed below, along with their respective antidotes, plus some parenthetical comments by the author.

ANTI-AUTHORITY: "Don't tell me." Antidote: "Follow the rules. They are usually right." (Indeed, many rules were written in blood.)

IMPULSIVITY: "Do something - quickly!" Antidote: "Not so fast. Think first." (Tony LeVier, the late and famous Lockheed experimental test pilot, who did the first flights of over fifty new airplanes, once said, "The very first thing to do in an aircraft emergency is nothing; analyze the situation before acting.")

INVULNERABILITY: "It won't happen to me." Antidote: "It could happen to me."

MACHO: "I can do it." (This one and that preceding are combined in the "young male invincibility syndrome" which, if not treated with frequent injections of common sense, can be terminal.) Antidote: "Taking chances is foolish." (And disaster plays no favorites among fools.)

RESIGNATION: "What's the use?" Antidote: "I'm not helpless. I can make a difference." (What's in the head makes the difference -- always.)

APPENDIX A TO CHAPTER ONE—THE PRECISION DUTCH ROLL MANEUVER

OBJECTIVE: To become familiar with the use of the flight controls for all three axes, while in a precise cross-controlled slip, to the degree that the pilot will develop the ability to make wing-down cross-wind landings with smoothness, accuracy, and ease.

PROCEDURE: Select an altitude that will allow an inadvertent spin entry and recovery with at least 2000 ft AGL remaining. (The maneuver is very slow and mild if done correctly. If the aircraft stalls in the slip the high wing can be expected to drop sharply and an immediate application of forward stick and neutral or opposite rudder, avoiding the use of ailerons, should effect recovery with little loss of altitude.)

If your airplane has a "Both Tanks" fuel selection, use that position to avoid the possibility of unporting a single, selected tank outlet. If "Both" is not available, the fuel tanks should be three-fourths or more full.

Establish approach airspeed and landing configuration, approach power, with about half flaps. Fly the maneuver level or very shallow descent. Do not get slower than Vs0 plus 5 knots, or a few knots above stall buffet, whichever is higher. But once you start the maneuver, the safety pilot can help warn you about too-slow speed. You should be looking forward at the nose of the airplane with respect to the reference point. You should have a CFI or a qualified safety pilot aboard

because you will be concentrating on looking forward during the maneuver.

Pick a distant object like a peak or cloud and point the nose below it. Then, if level and on speed, start the maneuver by slowly rolling the aircraft to achieve a steadily increasing bank, while at the same time using rudder control to hold the nose under the reference point. Add a little power to hold airspeed.

Use elevator control to counteract rudder effects and prevent the nose from rising. Your rudder inputs will also influence aileron effectiveness. Upon reaching 15-20 degrees of bank, or when you no longer have enough rudder available to hold heading, stop rolling briefly (five to ten seconds) and hold the nose steady below the reference point. Now, start rolling slowly in the opposite direction -- don't come off the top rudder too soon or you'll "scoop" out. Don't anticipate how much rudder you'll need -- use only what is needed, when you need it.

The technique of using only the rudder that you need when you need it is crucial to developing your cross-control skill. The measure of your success is doing what's needed to hold the nose steadily in place.

This maneuver can be very frustrating but keep plugging and you'll learn it well. Remember, you'll be in a slip, with the ball towards the low wing and using "high rudder" to hold the nose steady. This is the way the controls will be used in a wing-down cross-wind landing, except that in this practice maneuver your sole reference is to keep the

airplane aligned with respect to the reference point. You are simply learning control actions and interactions in the cross-control slip. In the actual wing down cross-wind landing you'll be varying the bank to stop the sidewise drift you see on the runway, while maintaining heading alignment with the rudder.

During your practice, add a little power at the start and don't get too slow or go below a previously designated altitude. Once you are comfortable and fairly smooth with the maneuver, you're ready to practice the real thing. Remember, rudder controls alignment/heading, ailerons control angle of bank for drift control, and elevators smoothly flare the pitch while resisting those pesky inputs from the cross-controlled ailerons and rudder.

Fly the maneuver both with and without power. Notice the effect that power has on rudder and elevator effectiveness, and how that will change a little when you change direction (left/right) of the slip. Notice that lowering landing flaps will reduce your rudder authority. IMPORTANT: Don't slip with flaps down if it is prohibited in your model of airplane. It's no fun to suddenly go nose-down ballistic with loss of elevator control!

APPENDIX B TO CHAPTER ONE - FLIGHT TRAINING LESSONS

Note: It is conceivable that you could use this as a guide to train yourself, or use this with an experienced bush pilot to help. However - and this is an important caveat - for safety and economy, these lessons should be given <u>only</u> by a fully qualified CFI.

SEQUENCE OF TRAINING MANEUVERS:

The general sequence of maneuvers should be to start with slow flight airwork and the Precision Dutch Roll to support takeoff and landing practice.

If the pilot is new to tail-dragging, it is recommended that the CFI <u>not</u> make the tail wheel sign-off until the FAR's are fully met, especially with regard to cross wind takeoffs and landings-to-full-stop. As an index, the CFI should require conditions that permit observation of good technique in a cross-wind ninety degree component equalling about twenty percent of Vs0.

Once the accuracy standards recommended in the first few pages of chapter one are achieved, there are maneuvers appropriate to different undercarriages and/or different phases of flight described in the following chapters that can be demonstrated and practiced. (See Caution at the end of this appendix.)

OBJECTIVES:

The flights should be started at or near maximum allowable gross landing weight. Only mild cross-winds should be used to start with, until the flight

instructor becomes familiar with the student's strengths and weaknesses.

The student should be able to make cross-wind short field takeoffs and landings, not only with the correct procedures but also with positive, smooth, on-speed control of the airplane. Soft/short field procedures with and without obstacle should be emphasized. The use of brakes for non-emergency directional control on the runway should be discouraged. Have the student make performance estimates for all takeoffs and landings.

Relying on brakes when operating on soft and/or ever-changing surfaces is poor operating technique and can result in some big surprise swerves and worse. Even on known hard surfaces it's a lousy habit.

At least one go-around from flare should be done each flight. The idea is to become skilled in slow flight evolutions at low altitude. This will lead to a healthier readiness to go-around when appropriate.

Introduce or review, as appropriate, the Precision Dutch Roll maneuver. Conduct the maneuver according to Appendix A, taking care to look out for adequate altitude and clearance with other air traffic. Do the maneuver with landing flaps retracted until the student has a good grasp of control usage.

Then, do the maneuver with flaps down at varying degrees up to maximum landing flaps, but only if permitted by the operating handbook. Emphasize positive, smooth control, using rudders only as needed to maintain heading. This maneuver should be well in hand before attempting any landings with

more than mild cross-winds (five knots or less, ninety degrees component). Caution: There are models which prohibit slips with flaps extended. DO NOT VIOLATE ANY SUCH PROHIBITION.

NOTE OF CAUTION: As explained later, off-airport conditions at any one place can vary considerably in a short period of time. Despite one's experience and skill, there is always inherent danger in all off-airport operations. For that very reason, I don't allow another pilot to be flying the airplane in an off-airport landing when I am aboard as the instructor (However, I did this as a Fish and Game check pilot because it was required.)

Rather, I will demonstrate landings from the front seat or right seat to full stops, usually two or three times. Then, if the student feels ready (of course, I will have already decided that he or she is ready), I'll get out of the plane and watch the student prove it.

A NOTE FOR THE FLIGHT INSTRUCTOR

There can be no compromises on the following evaluations. Knowing what to look for, the CFI can tell rather easily if the pilot at the controls has mastered good crosswind technique, even though the winds might be light and variable. One simply notes the pilot's flight control reactions, however small, to either heading drift or lateral drift when close to the runway. If, in the close-to-runway regime, the pilot being evaluated puts a wing down correcting for heading drift or uses rudder for lateral (sidewise) drift, he has not yet achieved good technique.

Similarly, if before or after the initial liftoff, the airplane starts a slight left turn due to increased P-factor and the pilot reacts by feeding in right aileron, rather than increasing right rudder, one can predict accurately that he will have problems with cross-wind, if any, during the subsequent landing.

As a CFI, I try to head off these problems by teaching two different flight regimes, right from the start with the primary student. In small general aviation airplanes and above about thirty feet AGL, it's a coordinated ball-in-the-center regime for normal flight. Closer to the runway, the pilot controls heading (alignment or cross-wind crab) with rudder, and keeps the wings level except to counteract cross-wind drift when using wing-down technique. As a result, the student pilot starts with what constitutes good technique for <u>any</u> relative direction of wind.

APPENDIX C TO CHAPTER ONE

VERIFYING BEST-ANGLE-OF-CLIMB (Vx) AND BEST-RATE-OF-CLIMB (Vy) AIRSPEEDS IN TERMS OF INDICATED AIRSPEED

Getting Ready

If you haven't already determined and recorded the indicated airspeed at which your airplane stalls with power at idle, at maximum allowable gross weight (MAGW), and with flaps down (Vs0) and flaps up (Vs1), you should do that prior to investigating indicated Vx and Vy.

Include in your load a qualified safety pilot to look out for other aircraft, and to time and record the results. The nearer to standard temperature-pressure altitudes it is, the better. However, you can make appropriate corrections to your recorded data. That is covered later in this appendix.

Pick a day when the wind is light to calm at all the levels you'll be flying. Rough air and up/down drafts could make your data useless. If there is a wind, fly your test climbs at ninety degrees to its direction.

Procedures

Caution: During these maneuvers, keep your airspeed well above the indicated power-off stall airspeed(Vs0 or Vs1), depending on flap position. Try to maintain wings level but never more than ten degrees of bank, safety permitting.

Have an outside air temperature gage available. If you don't have one installed, have the observer hold one in a cabin fresh air intake and record the temperature each time he records climb data. Temperature is important to determining density altitude, also correctly called "performance altitude."

Set your altimeter at 29.92 inches Hg. in order to record pressure altitudes alongside the temperatures (Re-set it before landing.) Record elapsed times in minutes and seconds from the nearest five hundred level after commencing the climb and then every five hundred feet thereafter. Doing every climb for a little over thousand feet will give you two sets of numbers for every climb. (For an example, see figure 1-7.)

IAS (mph/kt)	Pressure Altitude	Time (sec)	Time (min)	OAT (deg F)	Climb Rate (fpm)	Density Altitude	Notes (fuel,etc)
50	1500–2000	72	1.2	48°	416.7	1700	3/4' Full
50	2500	84	1.4	46°	357.1	2200	3/4 R Full

Landing Flaps Position (up/down) 20°
Weight at take-off: 1800#
Maximum Allowable Gross Weight 1900#

SAMPLE RECORDED DATA FOR Vx , Vy IAS
FIGURE 1-7

Trim the plane to hold it steady on your intended airspeed. Now, start recording pressure altitude and outside air temperature. Then, start the elapsed timer as you go through your lowest prime altitude. Get the first elapsed time and OAT when passing 500 feet above your start and a second set when 1000 feet above. Record a set of numbers for at least two

500 feet segments at each indicated airspeed. Return to the starting altitude slightly below your first recorded altitude and repeat the same procedure.

You can start the process with the flaps retracted to obtain Vy. First, use the published Vy airspeed. Configure your airplane for flaps-up best rate-of-climb. Make a minimum of two climbs for each run at any given airspeed. The idea is to verify your data. Whenever the second climb data is different than that for the first climb, do a third climb as a tie-breaker.

For each climb, use take-off power. Take care to avoid over-heating the engine. After the first run to gather data at published Vy, make your second run at an airspeed displaced by five kts/ mph. Try to bracket and nail down the indicated airspeed at which you get the highest rate of climb. Remember, do a minimum of two climbs at each airspeed for quality control of your data. (This will also illustrate the drudgery involved in much of test flying.) Record your estimated fuel remaining for each set of runs.

Next, for determining Vx, you will, of course, configure the plane in accord with the pilot's handbook procedures to clear an obstacle after take-off. Usually, but not always, that will be at or near twenty degrees flaps. Almost uniquely, the 150 HP Super Cub handbook recommends full flaps, climbing to obstacle clearance at 45 MPH, then accelerating to 75 MPH prior to retracting flaps. This

is due to the large reduction in ground roll and a consequent big advantage in starting the obstacle clearance climb earlier. Caution: forty-five miles per hour is too slow for safe flight control in gusty winds.

Vx is determined indirectly, by graph. If your airplane is a high performer, make your climb runs at ten kt/mph intervals. If your plane is not "super" at climbing, make your runs at five kt/mph intervals. You should gather the data at airspeeds between Vs0-(or Vs1)-plus-ten and Vy-plus-ten, depending on flap position. Then, convert the data into two smooth curves on graph paper. Using a "French curve" template will help in this task.

Plotting and Analyzing the Data.

Divide the number of feet climbed for each segment (500 ft) by the time in minutes (seconds divided by sixty) that the segment required. Enter that number in the last column entitled "climb rate" in feet-per-minute.

Using graph paper, first label the horizontal scale indicated airspeed scale, from zero to beyond the highest IAS tested. See figure 1-8, next page.)

Make the vertical axis the climb rate scale, in feet-per-minute from zero to a value higher than the highest rate-of-climb recorded. It is important that both the horizontal and vertical scales be continuous from zero through the highest figures encountered for airspeed and climb rate. Then, enter a point for

each line of airspeed/climb rate figures. Using a French-curve drawing tool, connect the points in a smooth line. Do a graph for flaps up and one for flaps at obstacle clearance take-off setting.

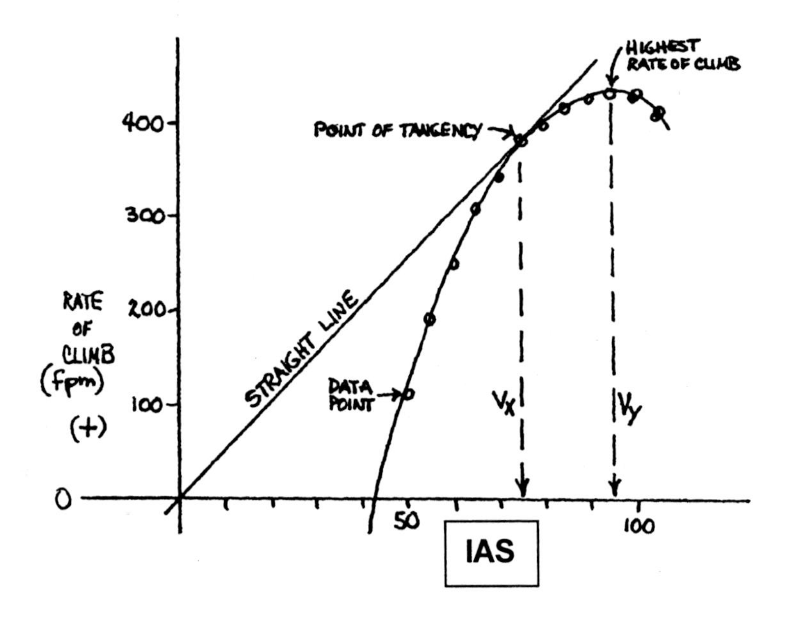

CLIMB CURVE

FIGURE 1-8

Now, for both graphs you have constructed, lay a straight edge from "zero-zero" to a point where it barely touches the curve that best connects the points. That "point of tangency" defines Vx as the airspeed directly below it on the horizontal axis. The highest point on the curve defines Vy as the airspeed directly below it on the indicated airspeed axis.

In some, but not all airplanes, having the flaps

down shortens the takeoff run so much that, even if flaps-up produces a higher angle of climb, having the flaps down gives you the overall advantage for obstacle clearance. In any event, follow the pilot's handbook for obstacle clearance takeoffs and climbs.

Corrections for altitude, temperature, and weight.

Using a temperature conversion scale and a pressure-density altitude conversion graph, you can convert your climb data to appropriate density altitudes. From that you can build your indicated airspeed Vx and Vy tables for different density altitudes. (See thumb rules, below.) If your pilot's operating handbook does not have a density altitude-temperature conversion chart, you can find one at National Weather Service stations. Also, you will find an excellent one in the FAA's Flight Training Handbook (AC 61-21A), at page 324. (AC 61-21A is for sale by the Superintendent of Documents, U. S. Government Printing Office, Washington. D.C. 20402. ASA, Inc.,1-800-272-2359, also sells it.)

If you find that your indicated airspeeds are within a percent or so of the handbook numbers for the same density altitudes, you can rely on your airspeed indications. Otherwise, you likely have either airspeed indicator or pitot-static system problems, or both, and should have the entire system tested.

Some General Thumb Rules Regarding Vx and Vy.

In the following thumb rules, we're dealing with

average, non-turbocharged, two-to-six place airplanes.

Vx *increases* about one percent for every 2000 feet increase in density altitude.

Vy *decreases* approximately one percent for every 1000 feet increase in density altitude.

Vx (flaps up) and Vy are equal at absolute ceiling. So, if your airplane has a big difference between Vx and Vy when at sea level, you likely have plenty of reserve power and good altitude capability.

Vx (flaps up) is appropriate for climb after departure if high terrain is a factor between take off and cruise altitude. Otherwise, Vy or even better: try a cruise-climb at Vy plus the difference between Vy and Vx (flaps up). That cruise-climb speed ought to be reduced about one percent per one thousand feet increase in altitude. Forward visibility and engine cooling should also be considered when choosing a cruise-climb airspeed.

When the airplane's gross weight is reduced below maximum allowable gross weight (MAGW), the entire climb performance curve shifts towards zero by about one knot per one hundred pounds reduction from MAGW. Therefore, Vx and Vy are reduced by about the same amount. You can and should verify this relationship simply by making two flights in the same conditions but with known differences of say, two hundred pounds total weight.

Remember: Vx is an "angle thing" and Vy is

“vertical speed thing.” Getting airborne early and accelerating in ground effect to Vx (flaps down) will sometimes beat Vx (flaps up), in short field obstacle clearance. Nevertheless, don’t guess; just follow the handbook procedures for your airplane.

CHAPTER TWO – PLANE, PROPELLER, ENGINE, AND UNDERCARRIAGE

Bush Pilots Fly Their Own – Considerations In Buying Your Airplane

Due to the hazards involved, it is a rare pilot who has permission to fly someone else's airplane in the wilderness. (The exceptions are government and air taxi/charter pilots.) Because of this the private pilot usually must provide his or her own airplane. This chapter is intended to help in your search for an airplane that suits your needs.

Looking At Equipment From A Bush Pilot's Perspective

This chapter is not a review of specific models. Rather, it is a review of how you can look at the equipment being considered for use: the airframe, powerplant/propeller, and the undercarriage. It looks at those things from the perspective of the bush pilot operating in marginal conditions. Where appropriate, specific airplane models and techniques are mentioned. The idea is to outline major factors you should consider prior to purchasing any airplane for operating in the bush.

Over-riding factors in selecting an airplane for off-airport flying, land and sea, include choosing a plane that has a slow takeoff and landing airspeed and has enough reserve power to enable a climb rate of at least 500 fpm while loaded to maximum allowable gross weight. If your plane can land and

takeoff at forty-five miles per hour, you will avoid much of the grief experienced by pilots who try to turn their stubby-winged speedsters with sixty five miles per hour liftoff speed into bush planes. They simply need too much room in which to operate. Moreover, the kinetic energy involved in hitting bumps in the off-airport setting increases with the square of the speed.

On the other hand, an airplane that lands much slower than about forty mph could have a problem with controllability in strong, gusty winds, due to its probable light wing loading. It depends to a degree on its design.

The Piper Super Cub with 150 hp is one of a few good examples of capable bush planes with excellent qualities for flying the bush once it is suitably modified with, among other things, larger wheels and back-up cables.

Good Visibility Is Essential

It is obvious yet needs emphasis: flying the bush is an intensely visual thing. You need to keep the windshield as unscratched and as clean as you can. When encountering reduced flight visibility and slowing for maneuverability, select a flap setting appropriate to the airspeed and for better visibility over the nose. (For this situation, the author uses landing approach speed with twenty to twenty-five degrees flaps.

The airplane should have an effective windshield

defrost system. F. Atlee Dodge, of Anchorage, Alaska, holds an excellent Supplementary Type Certificate (STC) for a windshield defrost and heater system modification. (He also holds STC's for other bush-type of modifications and is a good source of advice.) The defroster really pays off in cold weather when you load breathing pilot and passengers, or when you load game meat not yet cooled or frozen.

You'll also be grateful for a good defroster if you get caught in super-cooled rain falling from warm, overriding clouds above. (A later chapter discusses flight techniques for handling this potentially hazardous situation.)

If you wear glasses, keep them warm when outside the cabin in frigid weather. Otherwise, they'll be first to fog-up when you climb into the airplane. If you need glasses constantly, keep a second pair warm. This includes the situation where you are operating in bright sunlight from or over a snow or ice field and need to wear sunglasses to avoid snow blindness. If you use polarized lenses, do not wear them when operating a seaplane to or from water. If you do, you could have trouble discerning the surface – a very dangerous situation.

To help keep rain from rolling up into almost stationary drops on your windshield when you slow to land, and then catching the light to block your vision, here's what the author has found that works better than most commercial treatments. (You

should first test this or any method on a separate scrap piece of acrylic plastic windshield, just to make sure it won't craze the plastic.) Start by washing the windshield with plenty of water, your clean bare hands or a soft (clean!) cloth, and a good liquid dish washing soap, which has a wetting agent for thorough cleaning.

Next, after rinsing and drying with a soft cloth, apply a good quality spray-and-wipe furniture polish with a dry cloth. If it's raining, you can work under a tarp, but just being able to get it clean will help. This has worked well in some of the rainiest areas of coastal Alaska for float flying as well as for wheel landings on dark volcanic pebble beaches.

The author has tested commercial windshield polishes simultaneously with "Lemon Pledge"® furniture spray polish, by applying them in different areas of the same windshield. The furniture polish did better and lasted longer. Looking through raindrops catching the light against a dark landing area makes it difficult and dangerous to attempt a landing. This is so for wheels and floats.

Keep a collapsible bucket on board. Among other things, it will be handy for using clean creek water to wash the windshield as well as to melt early morning frost on the wings and tail feathers.

Never use window glass cleaner on windshields; it could craze the plastic. If you are negotiating to purchase a plane, take a look from the pilot's seat through the windshield toward and at angles to a

bright light, preferably sunlight. If the wndshield is crazed, cracked, or otherwise unacceptable, check the cost of replacement, including labor, before you settle on a price. Better yet, don't take possession until it's been replaced.

Protect Your Fuselage Bottom And Empennage

Next comes protection for your airplane's fuselage bottom and horizontal stabilizer leading edges against the ravages of small brush, gravel, and ice/snow chunks. If your plane has a fabric covering on its bottom, you should have the bottom "metallized" before doing any off-airport operations, to help prevent bottom damage. Bottom damage can compromise the integrity of your fabric covering, fuselage stringers, and basic fuselage box structure, thereby rendering your airplane unairworthy.

Another important benefit of installing a metal bottom is better access for inspection, maintenance, and repair. A possible drawback is slightly higher basic empty weight and maybe a small shift in the airplane's C.G. It's more than worth the cost and weight.

For horizontal stabilizer leading edges, you can use heavy-duty tape. Your mechanic will have a roll of this stuff. But take care here; the stabilizers on your airplane might not be tolerant of tape, for aerodynamic reasons. Check with your A&P on this important aspect.

Center-Of-Gravity Factors – Why the Tail Dragger Wins

Even with good visibility and a metal bottom for starters, the off-airport aviator must be equipped and ready for the weird things that airplanes will do on unfriendly surfaces, with and without unfriendly winds. In the case of the wheel plane a lot depends on whether it has a nose wheel or a tail wheel. Comparing the two types, the nose dragger is at a big disadvantage when operating from unprepared surfaces.

On rough or soft surfaces, with the center of gravity forward of the main gear, the nose strut and wheel take a beating and the propeller is exposed due to proximity to the surface, even if you load the plane to its aft C.G. limit. When the nose wheel encounters dips or soft spots the nose dragger wants to nose-dive. That's why in rough or soft area operations with a nose wheeler it is only a matter of time until you commit propeller "modification." Immediately following that, you get a chance to self-test for hazardous mental attitudes.

Nose wheel steering is handy but, in loading aft to protect your nose gear and prop, you lose a great deal of steering effectiveness. Furthermore, even with the C.G. at the aft limit, it is still forward of the main gear. As a result, the nose dragger will try to turn downhill on a side-wise tilted surface, like a beach or a hill contour strip. This downhill tendency creates more problems of finding the best wind for nose dragger landing on a sidewise-tilted surface. (More on those problems later.)

Nevertheless, with approved heavy-duty nose gear and appropriate large wheels, the large

"trikes" can do well when flown within limits more restrictive than those for tail draggers. For example, the author has landed big-wheeled U-206's on shallow gravel bars and on wide mountain ridges which had been flattened by wind and were fairly straight with hard packed dirt and gravel. With decent wind, these were not chores because they were well within the nose dragger's limits and there was acceptable room for the unexpected. (One thing about ridges: you usually don't have landing approach and departure obstacles.)

In contrast to the nose wheel airplane, the tail dragger is much more adaptable to off-airport operations.

The tail dragger does tend to turn uphill because its C.G. is behind the main wheels. Whereas the nose dragger's forward C.G. promotes prop-surface contact, the tail dragger's aft center of gravity counteracts that tendency. Full back stick combined with high power will reinforce this feature in situations covered later. In sum, with good pilot techniques, the tail dragger can handle much softer and rougher operating surfaces than those tolerated by the tricycle wheel plane.

Engines and Propellers

When purchasing an airplane you should look at the engine/propeller combination carefully. If high density-altitude operations are planned for a four-place or bigger craft, then a turbo-charged or normalized engine will likely be your best bet for performance. For maximum life from the turbo-charger you should observe strictly the

turbine cool-down time. That can be inconvenient at times, but you have to live with it, or expect short-lived turbine bearings (coking up) and oil seal (drying out).

With traditional flat pitch designs, two-bladed propellers can be expected to improve takeoff and climb performance by about ten to fifteen percent compared to three-bladed props. However, their longer blades are closer to the surface and, with faster tips at the same rpm, are noisier. This, of course, is a negative factor in many localities. In such cases, you'll want to be circumspect and reduce to climb power rpm's as soon as it's safe to do so after takeoff. This is not a big problem with the smaller, four-cylinder engines.

One of the greatest developments for the Super Cub was Rodger M. Borer's STC for the longer, flat-pitched McCauley propeller in 1967. If you are accustomed to a cruise prop (such as a 7556) and hang a Borer prop, like an 8241 (82-inch diameter with a 41-inch pitch), on your Super Cub, you'll be amazed by the improved takeoff and climb performance. What you give up is less efficiency at cruise speeds.

If you're planning to put your newly purchased tail dragger seaplane on <u>standard</u> size wheels, and it has one of those long two-bladed props, that propeller will be too close to the surface on a wheel landing. For this intolerable situation, install larger tires or get an additional, smaller prop. Utility as well as cost ought to drive your decision. If you're going to use the plane off-airport, installing large (25" to 32") tundra tires will be your best option on both counts.

There is a significant consideration when you install bigger diameter tires but do not install larger diameter wheel brake rotors. The brakes will have even less mechanical advantage because of the greater radius of the big tires. That's not all bad -- you'll want to use minimal braking on soft surfaces anyway. Some pilots install boosted brakes.

If you are contemplating using a small two-place aircraft, the plane might not have the engine power you need for float flying. The reason is the big increase in basic empty weight due to the floats. There are many STC's for upgrading particular airplane models with specific engine/propeller combinations. Talk to experienced bush pilots and mechanics about this. Don't forget, higher power engines cost more to acquire, operate, and overhaul.

If you decide to buy a used engine, before you close the deal, have an engine shop give it an inspection and test run, followed by an oil analysis. Be leery of overhauled cylinders, especially chromed, that have been welded. New and first-run cylinders are best by far.

For normally aspirated engines, an overall fuel-usage planning factor is to multiply the horsepower by 0.06 to obtain approximate gals/hr at cruise power. For flight planning, use the performance tables. Actual cruise power setting can make a big difference in fuel consumption.

If you plan to have the engine "souped-up" with higher compression, you will need more cranking power for engine starting. This is especially

important for float operations. The battery and cranking system already installed might not be adequate for the bigger engine. The real test comes when the battery is cold and the engine is warm.

Having spent a few nine-month winters in the interior Alaska Range hand-propping a 160 hp Super Cub with standard starter, the author can attest to the value of a geared starter drive with appropriate starter ring gear. (Hand-propping does build character.) Oh yes, the higher compression engine will likely require higher minimum octane fuel.

Undercarriage

During the warm season, large tundra tires will give you capabilities for off-airport operations and better prop-surface clearance. That's so, however, you could need wheel-skis for cycling between low altitude civilized strips and the high snow fields. Wheel-skis usually require use of standard-size tires.

If the snow is hard-packed or thin over smooth ice areas, the large tundra tires work very well. However, if there's any doubt or if the weather is frisky, go with wheel-skis or replacement skis until you're sure. The author has used both huge tires and skis on snow/ice fields and found the big tires much easier on the plane in shallow but suitable ice/snow mixes of summertime. If you've got snow at home base and in the hills, you are better off with replacement skis because they will give you the best flotation.

Some experienced seaplane pilots have even

operated float planes successfully on high snow fields from bases on water. There is no reason it wouldn't work well for delivery of loads to a sloping field when conditions allow landing up slope and departure down slope. A number of pilots simply go from skis in the winter to floats in the summer.

Float Planes

If you intend to buy a float plane, you should become well-informed on float construction and features. There are perfectly legal airplane/float combinations that might be marginal in their capabilities to operate in the water conditions in your locality and/or at the weights you plan. The biggest allowable floats may not always be the best either, because of the weight/drag penalty. What you want is a set of floats big enough to support the maximum weight with adequate flotation reserve for the kind of water conditions in which you'll operate. That will take some research on your part, and some talking with disinterested parties on the pros and cons of different options.

There are some floats with relatively deep-V cross-section and sturdy build that are very useful in salt water or windy fresh water operations. The deep-V floats cut into waves with less impact shock transmitted to the fuselage and its contents. The shallow-V floats can handle calm water best with average airplane/powerplant combinations. With the shallow-V floats, you may have to use the "porpoise" technique to get up on the step promptly with full loads but that's not a disadvantage in itself, just different. On the other

hand, the shallow-V float-equipped plane takes more beating in rough water. The author favors deep-V floats.

Traditional twin float construction uses N-struts directly connecting the fuselage to two floats, which are attached to each other by horizontal spreader bars. Adjustable "flying" (aerodynamic) wires, used in X's, help maintain the dimensions within the entire structure. During your first ride in the average float plane you literally are in for a shock from the forces that can be transmitted to the fuselage during normal operations.

Floats that are mounted on flexible main landing gear struts can be expected to transmit less apparent shock to the fuselage. But with these floats, you will want to check frequently the main landing gear boxes, where the main gear struts are attached within the fuselage -- as if you were operating the plane with wheels on rough surfaces. All gear boxes were not created equal. With this kind of float mounting, you should talk to an A&P and learn the gear box inspection procedures (where to look, how it should look, and how damage is evidenced).

With any kind of floats, ensure that your floats have propeller spray shields installed and also put on a propeller surface protectorant to help a little in protecting your propeller. Water, being incompressible like gravel, can erode and damage your prop almost as quickly as gravel. Good operating procedures, like avoiding unnecessarily high power settings and keeping the elevators where they belong, will help you avoid prop-damaging spray.

There is always the temptation to replace your standard size floats with larger approved floats in order to increase the useful load. When considering this option, note that the center of flotation could be relocated with respect to the airplane. Consequently, your optimum takeoff technique and the best pitch attitudes for takeoff and landing could be different than that to which you're accustomed for standard floats. Usually, aerodynamic drag will increase and the aircraft's performance will decrease.

The distribution of aerodynamic sail area likely will change, with an increase forward of the CG. That would rob your plane of some positive lateral dynamic stability—that is, if you push and release a rudder while airborne, the nose will be less prone to resume its original alignment with the airstream. It's best to have a pilot familiar with the new combination give you a check-out flight.

Finally, with larger floats, your plane could have reduced positive dynamic stability when touching down on the water. This requires less tolerance for cross-wind drift and perhaps slower water contact speeds for landing. These are things you should check on before switching to larger floats. With any kinds of floats, avoid fast, flat touchdowns and cross-drift to minimize hazardous lateral instability.

Amphibious Float Planes

If you are considering operating an amphibious float plane, here are some factors to consider.

First, the retractable landing gear system adds weight, cutting down useful load even more than

"straight" floats.

Second, there is consequently great temptation for operators to overload the amphib. Overloading invites long takeoff runs on spacious water areas and an engine that could have been overheated and perhaps damaged internally. If the plane has a constant speed propeller, it will be tuned to takeoff at maximum rpm in an effort to extract every horse from the engine. That's okay, but governors and tachometers sometimes drift off, in one direction or another. Have the engine oil analyzed, the engine bore-scoped, and otherwise inspected thoroughly prior to buying a used amphibious floatplane.

Third, most amphibious floats have complex landing gear retraction-extension systems, some models being much more complex than others. They come in two basic categories: those with <u>separate</u> control/actuating and position-indicating systems, and those with inter-dependence among the subsystems. With the latter there is a higher probability that if one subsystem fails it could cause a chain reaction—indeed sporty if you're engaged in takeoff or landing. You should understand the systems, regardless of how complex, all of their failure modes, and the required pilot actions for different failure modes and scenarios.

Fourth on our list of amphibious float peculiarities is the pronounced "design to minimum weight" philosophy. This is certainly understandable, as well as evident, in the small landing gear size. That characteristic doesn't allow

much latitude for operating in the wheels-down mode on other than civilized airstrips. Even in the best of conditions, you should keep serviced and inspect daily the brakes, wheels, tires, and retract/extend systems for leaks and damage.

Fifth, most amphibious nose struts are not the strongest. Be very reluctant to taxi wheels-down up onto a beach. The author strictly avoids taking an amphibious floatplane wheels-down from water-borne up any incline that is not smooth, firm, and shallow, like a man-made ramp. For almost all beaches, treat the amphibian like a “straight” seaplane and you’ll avoid nose strut problems.

Sixth is the fact that amphibious floatplanes are, to varying degrees, “draggy” on water and in the air, plus they fly a bit wierd in the lateral stability department. Floats in general add a lot of sail area forward of the C.G. and some amphibs add even more. That results in even less positive lateral dynamic stability than with standard floats. For instance, if you feed in rudder to put the ball off-center while airborne, then release rudder, the ball might very well stay where you first put it. Not a bother unless you’re flying for hours in rough air and like to keep the ball somewhere near center for performance and comfort.

The amphibian, like the tricycle gear wheel plane, has its center of gravity forward of the main (aft) landing gear, but more so. When you touch down on the main wheels during landing, the sail area

forward of the wheels is even greater than that forward of the C.G. Therefore, in a crosswind, the amphibious float plane will arc downwind, instead of upwind. If you are on loose gravel, it gets even more sporty and you may have to "crab" into the wind early on rollout to stay on the runway. That's right, you might have to do some "gravel-sliding" for a short distance in order to stay on the strip. To minimize that, you'll want to get rid of lift early by going to "all four" and raising flaps to get more weight on the wheels for stabilizing things. (Make sure it's the flaps you're raising!)

Driving an amphib on land is a little like driving a coffee table, with two wheels on the rear in fixed alignment and the front two almost-free castering. It's weird but fun. Because of these characteristics as well as the high C.G. on amphibs with gear down, expand your runway cross wind operating envelope with caution.

On the water in the displacement mode, the amphibious floatplane tends to nose down with the least provocation. Accordingly, you should take care to keep the water taxi speed down and the elevators up for most displacement mode operations, depending, of course, on the wind situation.

Finally, consider the dread of all amphib drivers: landing on a runway with wheels up or touching down on water with gear down. The runway can damage the floats and your pride while the water will put you vertical or on your back promptly.

Always use a check list that's posted in sight, and do the ditty about, "I'm landing on water, so my

gear should be up," and vice versa for land operations, and touch the landing gear control, visually verifying lever position and position indicator(s). Convex mirrors on the lift struts and/or on the inboard side of the floats will help you see the wheels when they're down. These procedures will help prevent what some people call the "inevitable" accident. Believe me, it is not inevitable if one uses good procedures without fail.

Despite all its peculiarities, the amphibious floatplane is fun to fly and has impressive operational flexibility. It could be just what you need.

We will cover some operational techniques for float operations later but, for now, the author hopes he has convinced you that you should do everything you can to become well-informed on float lore before you make decisions about the floats you need.

Ski Planes

If you are happy flying only on wheels or simply have no interest in water flying (although I can't imagine a more delightful type of flying), you might yet want to try ski flying, which can be done with the land plane pilot rating.

With skis, as with floats, you should always be aware of how much "flotation" you need for the surface you'll be on. The replacement skis give you the most flotation, for the deep, fluffy snow. Make sure your skis have the required runners on the bottom to help with directional stability. If you're flying from sticky snow during warm spells, it

helps reduce surface friction to be at the minimum required number of runners. Some ski "sleeves" made of three to four millimeter thick plastic can give you temporary relief for a sticky snow takeoff.

For durable slipperiness, most operators have gone to using a Teflon layer on the ski bottom. Even so, take care not to go below the minimum number of runners for your skis. You need good edging for directional stability. Otherwise, you could end up going a little "sideways" in a turn because you removed too many. The next thing that could happen is to side-swipe a hard bump with the outside edge of your leading ski and promptly destroy the landing gear on that side. Runners are also valuable as sacrificial metal when the snow on a gravel runway turns out to be thinner than expected or reported.

There is at least one ski modification on the market that provides a retractable spike actuated by the wheel brake system. It is likely effective especially on hard, icy surfaces, but the author has no experience with this. It's certainly worth looking into. It should be a safety boon to operating on icy surfaces.

In the meantime, get ready for surface ice by setting your idle rpm at its legal minimum. Additionally, you can use carburetor heat and alternate between magnetos to keep your taxi speed slow. You can also stay ready to shut down, step out of the plane, and pull the craft into a turn in the desirable direction in order to avoid a crunch situation. For example, in a Cub, you'd be ready to make a turn to the right, and therefore allow the

most room to the right; all because you have to egress to the right. For the 180/185, it would be a left turn. (Caution, all this is still <u>not</u> adequate for taxiing in tight quarters.)

Practice the quick egress maneuver, with the engine shut down, before you are in the actual situation. You will find that you could use some non-skid surface applied to the top of your skis in order to make this a safe maneuver. (Right about now, the retractable spike is looking even better.)

Without the spike, if you can avoid icy surfaces, do so. In any event, do your utmost to avoid taxiing <u>downwind</u> – especially in tight quarters. It will be the worst of a bad situation. That's due to the lack of rudder control, which can be countered with propeller blast against rudder to regain control. In turn, this speeds up the airplane. Then things get downright sporty because you are now "in chains."

To avoid all of the above, some pilots will rig a couple of turns of heavy rope around each ski in order to create equal surface drag and be able to use power without accelerating. This works very well when done carefully. You should shut down before rigging anything on the skis, and make sure that nothing can get tangled in the propeller when you re-start for taxi.

Two developments that have been around a long time are the penetration ski and the wheel-ski. These skis will usually be certified for use with standard size tires. The penetration skis mount on the main and tail wheels. They are larger than the simple replacement skis and each ski has an opening for its respective wheel to penetrate an

inch or two. Penetration skis are usually rugged and very useful where you need maneuverability. In this regard, check out the newly approved light weight but strong penetration tail wheel ski by Burl Rogers of Chugiak, Alaska.

On the other hand, the wheel-ski can be mounted only on the main wheels with no change to the tail wheel except in certain situations. The skis are raised or lowered by hydraulics, hand-pumped or electrical pump, depending on the model. Operation manuals for wheel-skis are scarce. For that reason, we include the following tips.

There is less aerodynamic drag with the skis down, so you'll want to climb and cruise in that configuration. Skis-up configuration makes for a "speed brake" of sorts and allows one to use higher power during descent to help keep the engine warm.

While taxiing on wheel-skis, there are times when you'll need to retract a ski and allow the wheel on that side to provide the drag necessary for a sharp turn. Simply hold the wheel brake on the side <u>opposite</u> the direction of turn to hold its ski down while pumping the other ski up. When you feel the inside wheel settle off of the retracting ski, you're ready to start the turn with a batch of power, full rudder, and enough forward stick (yoke) to get the tail off the surface or at least "light" on the surface. (Take care not to nose-up.)

After the turn, pump for skis-down while moving slowly to allow the up-ski to slip under the inside wheel. This really beats the same situation with

replacement skis where, depending on the situation, you may have to shut down, get the passengers out, push the tail around, re-embark, and re-start. Or you can anticipate the need for the turn and rig a few drag-producing turns of rope to go under the inside ski when you need it. In this regard, with experience, you will likely develop your own rig for this situation. Take care not to rely on a rig operated by anyone other than you, the pilot, and is retrievable from the cockpit. It should have a simple, positive release and, of course, be clear of the propeller, hydraulic lines, and rigging wires.

With wheel-skis, prior to each landing you should go through a ditty similar to that of the amphibious water pilot, “I’m landing on snow; my skis should be down,” or, “I’m landing on gravel (concrete); my skis should be up.” And then you check the controls, indicators (if any), and look at the skis directly. Strut mounted convex mirrors can help you see the “blind-side” ski, if there is a blind side.

Having operated a U-206 on a set of French-made nose dragger wheel-skis, we found the rig to be useful. The nose wheel damper gets an assist from a nose strut-mounted compressed air damper. We had to be careful to maintain correct damper pressure or get big vibrations. Doing that was a challenge at temperatures below about -20 deg F. Otherwise, it was a highly satisfactory system with good directional control.

Whether or not you should replace the tail wheel with the small replacement tail ski depends on the kind of surface on which you’ll operate. For deep, fluffy or wet, sticky snow the small replacement tail

ski does well for maximum flotation with minimum drag. Take care that a qualified A&P rigs the tail ski.

In many situations, leaving the tail wheel installed can be useful as a brake by holding up-elevator, with or without power, depending on your speed. Moreover, if you need positive tail wheel steering, stay with the tail wheel. It can make a difference where directional precision is a must.

A final note about putting your airplane on skis: you should know with certainty that the skis are properly rigged. This includes how the skis are mounted and the exact lengths and angles of the bungees and safety cables.

The author once met a pilot who had recently survived a very sporty incident resulting from bad ski rigging. With newly installed skis on his Cessna 185, he had just taken off, headed for Fairbanks. Shortly after leveling off at about 4000 feet AGL and as he picked up cruise speed, the airplane started a steady, irresistible nosing down.

Realizing he couldn't get the nose up and was too steep to roll over at that point, he chopped the power and did an "outside split-S" to a climb. He then rolled upright, stayed slow, and made an emergency landing on a frozen river.

It turned out that the ski-toe bungees and safety cables were too long. So, when he leveled off and nosed over to cruise speed, the air stream got on top of the skis and turned them into nose-down control surfaces. Fortunately, the pilot had the altitude, ability, and presence of mind to do what

was needed. After the split-S, the airplane was kept slow with the ski toes high enough for reasonable control. But the aircraft was damaged by the escape maneuver. The moral of the story: make sure your skis are rigged precisely "by the book."

Wheels

For the wheel airplane, the large tires are necessary for almost all off-airport operations. Having flown the Super Cub on a wide variety of "tundra" tires, in the author's opinion, they should not be larger than about 33 inches in diameter or wider than about eleven inches. And having tested them up to 36 inches in diameter and to about sixteen inches in width, the author found that those with the 36-inch diameter and also the wider tires interfered aerodynamically with elevator and rudder control in a slip. Also, they showed slight gyroscopic action and there were large spin-up moments when touching down on a gripping runway. Even with larger brake rotors, the largest size made braking marginal.

Nevertheless, the 36-inchers are effective in stable snow if you carry about six psi air pressure in them. Overall, the author would not recommend their use because of their interference with flight control effectiveness in the slip maneuver.

For the average tundra tire, about six to eight psi air pressure will work, depending on the model. But they might not stay on the rims if you make sharp turns on rough surfaces or operate on steep beaches or steep sidewise-tilted strips. For that, the most satisfactory are the 29X11X10 tires that are used for the Cessna Caravan. We put them on the

Super Cub and carried six to eight psi air pressure. They have stiff side-walls that help keep them on the rims on steep beaches. However, they come with grooved treads, and that helps them pick up gravel and mud more readily than the dimpled smoothies. So, if you use them without filing off the tread get ready for a very dirty airplane on muddy surfaces. We never did file off the treads and suffered for it only a little.

The 29-inch diameter mains are also a good bet for the larger six-cylinder tail draggers. (I've used them on Cessna 180's and 185's') Of course, big tires create a lot of aerodynamic drag. Therefore, if cruise performance is a major concern -- and it certainly can be in some situations -- you might have to limit the large tail dragger's off-airport utility by using intermediate-size main tires.

A major limiting factor in off-airport operations with wheel planes is the tail wheel and its supporting structure. Here's where you simply have to work closely with your A&P to ensure that the tail end is well-equipped, sound, and legal; the three go together.

For example, you could make your tail wheel structure extremely strong but illegal with a "beefing up." Subsequently, if you hit a damaging object in off-airport operations, the wheel and supporting leaf springs might look and actually be okay. But they would have passed the force on up, well into the fuselage structure, doing serious damage. To prevent this sort of thing, approved modifications are subjected to stress analysis prior to official approval. That's why it's important to

work closely with your A&P on any modifications you have in mind.

Anyone in the bush flying business will agree that the standard Super Cub tail wheel leaf springs are not strong enough. However, Pawnee leaf springs make a welcome difference, and work well with the Scott eight-inch tire, or the “high flotation” tail wheel that’s now on the market. It would have been nice to have had the latter when working the volcanic beaches of the lower Alaska Peninsula, but we got used to having the tail wheel disappear into the black "fluffy" as the airplane slowed down.

You cannot clean and grease the tail wheel fittings too much when you are working regularly off-airport. Further, you should take the tail wheel apart frequently for cleaning, including the inner tube. Take care to let the air out of the tire <u>before</u> you loosen any axle nut—the wheel could be fractured and held together against internal pressures by only that nut! Have your A&P check you out on the safe care of wheels and brakes, including what to look for when operating in rough conditions and inspecting for damage on pre- and post-flight inspections.

Spin-up of some types of large main wheels tends to get the inflation valve stem out of line with the hole provided in the wheel casting. You’ll have to disassemble the mains periodically, clean out the inside of the tire and reposition the valve stem properly before re-assembling and inflating.

The author once had an interesting experience following a fourteen-hour day of doing what is called “stream marking.” Using a big-wheeled

Super Cub, he had spent the day landing and erecting commercial salmon fishing regulatory signs on a number of fluffy, volcanic-type of beaches. Although he had cleaned and lubed the wheels the day before, when he returned to land on the concrete runway at Cold Bay, a thrill was waiting. After making the usual wheels touch down, he allowed the tail wheel to settle onto the runway. The aircraft quickly veered to the right, towards some aircraft parked on the main ramp. The author countered with left brake, power, and left-forward stick. That worked, fortunately, and we returned to the runway under a semblance of control, and luckily without hitting the runway lights.

Disassembly of the tail wheel swivel mechanism showed it was full of fine, volcanic sand. It probably had stuck to the right when I made my one-eighty to the right for the last beach takeoff. Counter-braking with appropriate power straightened the wheel for the remainder of the roll out. It was yet another lesson about what can happen when operating in uncivilized places.

CHAPTER THREE

PLANNING YOUR WILDERNESS FLIGHTS

There are a number of things that affect flight planning, including that for the bush environment. Among the most important are: time of day, fuel, route, terrain, weather, communications/navigation equipment, and type of destination.

Time of Day

Most of the time, you will do best to fly as early in daylight as you can, even if that means a preflight inspection by flashlight at "zero-dark hundred." Basically, an early takeoff will help you avoid getting behind schedule and, as a result, being tempted or possibly forced to fly after dark over hazardous terrain though you or your airplane are neither equipped nor ready for flight with the owls.

The need for a first-light takeoff is increased when you're faced with a shuttle-in or shuttle-out situation. In short, banker's hours don't work for the bush pilot. So, if you're a "night person" consider cranking your biological clock back to that of an early riser.

Ocean Beaches and Timing

For an ocean beach you'll want to time your takeoff so as to arrive over the beach two or three

hours after high tide. Steep beaches could require more delay, especially with a narrow tidal range. In addition, the steep beach could have shorter useful periods for landing and takeoff. Because of the steeper beach angle, the width of useable beach is less for the same distance down that the tide level moves. (As you can see in figure 3-1, with equal tidal ebb, steeper beach A gets less increase in landing area width than does shallow beach B.)

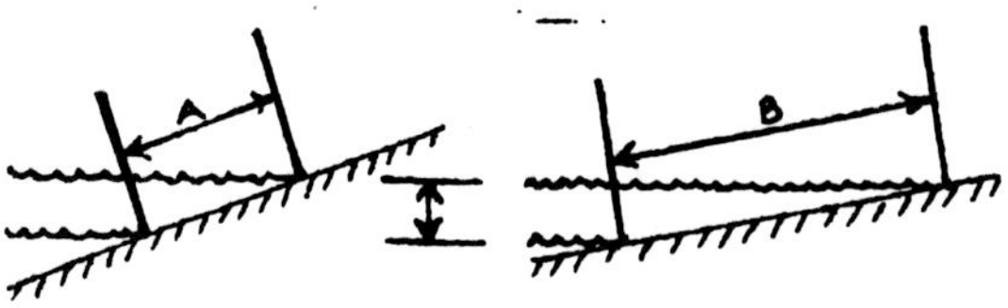

STEEPNESS VS. BEACH LANDING AREA WIDTH
FIGURE 3-1

Tide table booklets, which can be found at any commercial harbor or seaplane fuel dealers, are very useful in estimating what the tide might be at any time, even if your beach isn't listed. Using a chart that shows the listed tide table reference points and your destination, you can make estimates of what's happening where and when. Strong onshore winds will make both high and low tides higher than predicted, and strong offshore winds will make them lower.

The Lagoon Effect

If you're planning to use an ocean lagoon or river mouth or their beaches, there will be different times and magnitudes of tidal highs and lows, depending on where and how restricted the lagoon entrance or

river mouth is. Periods and ranges of low and high water could be different than you expect. From experience, you will learn how much “lagoon effect” exists with any particular lagoon or river mouth.

CAUTION: A real hazard when using lagoons and other protected waters is that water birds use some of them for food and shelter and for mass takeoffs and landings. You should look for this before choosing lagoons and lakes or their beaches for your base. It’s no fun cleaning feathers and avian parts from your propeller, windshield, and airfoils.

Lake Beaches

The acceptability of lake beaches usually depends on the season. That is, during spring run-off, the beaches will not only be very narrow due to high water, they could also be fouled with logs and debris brought to the lake by rivers and creeks. If areas of a given beach tend to line up with the strong winds, those areas will likely be the cleanest and most suitable for use. If not, then you might have to wait until later in the season when the lake level drops.

Before you depart, check carefully on the legal restrictions that might exist for using the destination lake. Before landing, look for and respect, boats and houses on and around the lake.

Waiting For Critical Weather Reports

If you must take a route through a "one-shot" pass, by definition, you will not be able to turn around safely once you enter that pass. So, you will want to get as much weather information as possible prior to takeoff. Certainly, you need weather information about the other end of the pass before entering the pass. This could mean that you must wait for a nearby surface weather report and the all-important PIREPs if it's reasonable to expect them. (More on planning for passes later in this chapter.)

CAUTION: **Most passes have elevations which are higher in the middle than at the entrance or exit, and therefore the middle is more likely to have clouds closer to the ground than do the ends of the pass. Avoid a one-shot pass that is higher in the middle whenever there's a chance of precipitation or low clouds in the area.**

Be alert for those locales close to bodies of water and which therefore are hosts to frequent early morning radiation fog or the on-shore flow of advection fog. Places having recently had rain and which are having a clear night could be foggy.

Fuel Planning

Because of the frequent unexpected events in bush flying, the author has found that a minimum of one hour reserve fuel is best. You may need <u>more</u> for long, rear-end numbing flights. The longer your

flight the greater the chances are that weather might force you to fly different routes. Some aircraft could have inaccurate fuel quantity gages. Therefore, it is very useful to back up your gages with estimates, including the fuel consumed for landings and takeoffs. Estimates based on the pilot's operating handbook can be a good start.

At the first opportunity, it will be useful if you calibrate a wooden dipstick for getting accurate readings of fuel remaining. Don't use a plastic dipstick. It can hold a dangerous electric charge. Starting with empty fuel tanks, and the fuel selector not on "Both," fill a tank five gallons at a time, marking the dipstick at each five gallon level. For extended-range wing tanks, fill from the outboard filler, giving each five gallons time to run into the inboard areas, depending on the model.

Never use your reserve fuel for anything less than emergencies or unexpected, important events. Do not use reserve fuel for unplanned, unnecessary actions, possibly putting you and your passengers "up a creek" for no good reason. Reserve fuel is for emergencies only -- it's a good rule.

The routine act of switching fuel tanks becomes a heart-stopper if you are over inhospitable terrain or wild water and the engine quits after you switch. Plan your fuel tank changes for the best locations and allow the newly selected tank to prove itself for a few minutes before you go into a less tolerant situation.

Using Automotive Fuels

What about the use of automotive fuels? To use unleaded auto gas, first get the STC that encourages occasional use of leaded aviation fuel for the replenishment of lead for valve lubrication. Follow carefully the instructions accompanying the STC.

Make sure your carburetor has a metal float instead of a plastic one. Also, ensure that any hoses, o-rings, seals, or gaskets that might come in contact with your fuel are impervious to ethanol. Ethanol, which causes many materials to swell and disintegrate, should never be in the auto fuel that you use in your aircraft. Gasoline mixed with alcohol is not approved by the FAA for internal combustion aircraft engines.

Engine experts will tell you to use one hundred percent aviation gasoline, with TCP as required, for engine break-in and the first fifty hours. After that, at least one engine expert recommends mixing ten percent aviation leaded gasoline with ninety percent auto gas if you have the STC. That will ensure adequate break-in of the valves and their continued lead lubrication, for which they are designed.

Testing Fuel For Alcohol Content

Here's a simple test for the presence of alcohol in your fuel. First, put about an ounce of <u>water</u> in a clear glass or plastic container with an available tight cap , set it level, and carefully mark the water

level with a grease pencil (china marker). (About ten percent of the available volume is best.)

Second, complete filling the container with a sample of your fuel and tighten the cap on it. Third, shake the mixture and place it on a level surface. Allow it to settle a few minutes and then observe the water level. If there is alcohol in the fuel, a small amount of it will have combined with the water and raised the water level above the mark you made in the first step. However, this test will not tell you what kind of alcohol is in the fuel. But, it likely will be ethanol, which is commonly used as a cheap oxygenate for auto fuel.

Auto Fuel and Vapor Lock

Automotive fuel made for use during winter has a higher volatility than fuel blended for use in summer. Higher volatility means higher probability of an engine-stopping fuel vapor lock. You can make that probability even worse by using winter auto fuel in the summer.

Both summer and winter auto fuel are much more volatile than aviation gasoline. Winter auto fuel is the most volatile. Accordingly, if you use auto fuel and want to reduce the possibility of vapor lock, avoid using auto fuel that you bought in the winter when you go flying in the summer. Additionally, you should ensure that your fuel has a smooth flow through the fuel supply lines to the engine, especially downstream of the "gascolator," also called the strainer bowl.

As an A&P (IA) making annual inspections, the author has seen airplanes with the auto fuel STC, that have a sharp ninety-degree fuel fitting downstream of the gascolator. That fitting can be replaced with an approved fuel line that causes the fuel to flow in a smooth barrel-roll/loop to the left, coming out ninety-degrees to the right and headed straight for the carburetor. This works best for installations where the gascolator is to the left of, and slightly behind, the carburetor inlet. (See figure 3-2, below.)

The same idea can be adapted to any installation, taking care to have the line stabilized and clear of, and insulated from, any heat source. The smoother flow reduces the probability of vapor lock due to fuel turbulence.

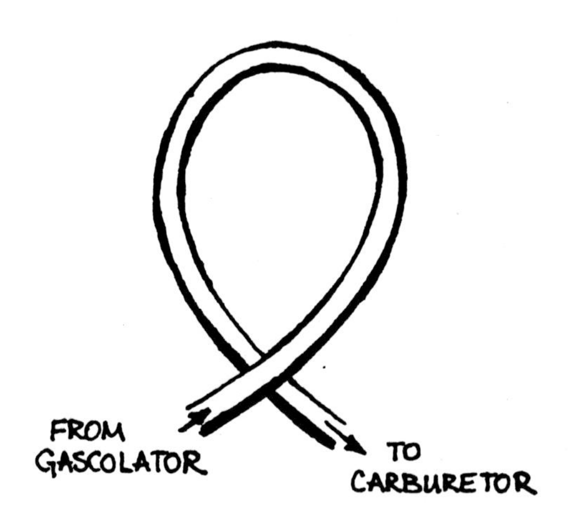

FUEL LINE ROUTING
FIGURE 3-2

Auto Fuel, Oxygenates, and TCP

The Clean Air Act requires the use of oxygenates in automotive fuel primarily during the winter in many geographic areas in order to reduce undesirable auto emissions. Three different additives are commonly used for oxygenates. They are ethanol (made from grains), MTBE (methyl-tertiary-butyl-ether), and ETBE (ethyl tertiary butyl ether). MTBE is being questioned, ecologically. As mentioned above, ethanol is not suitable for aircraft. However, ethanol is a common oxygenate. Before you buy auto fuel for your airplane, check with the fuel merchant. All dealers are required to keep records on whether their fuel contains oxygenate and what kind of oxygenate it contains.

Rather than buy auto fuel blended with ethanol, the author strongly recommends that you purchase aviation gas that meets your octane requirements. If your engine is designed for eighty octane and you can find only one hundred low lead gas, then by all means get the one hundred octane low lead <u>and</u> add TCP (or its equivalent) in compliance with its accompanying directions.

TCP (tri-cresyl phosphate) in the right blend will keep lead in vapor form much longer during the combustion exhaust process and thus help limit lead deposits on your spark plugs. Most, but not all, of the lead vapor gets time to get out through the exhaust port. The author used TCP for years in fueling Beavers and Super Cubs with both “Blue” and even “Green” gasolines, and the plugs stayed

fairly clean. On the other hand I've seen operators use the one hundred low lead gas without TCP in eighty octane-rated engines and experience "surprise power-out glides" due to plug fouling. TCP is being replaced by a similar additive more friendly to the environment.

Pre-positioning Fuel

In planning your flights you will find that pre-positioning of fuel will allow you, in many situations, to do the job without coming close to your maximum allowable gross weight. Being able to shuttle in or out and consolidate reduced loads at bigger strips opens up smaller areas for safe off-airport operations. For pre-positioning, use sealed fuel containers (new is best), tag them with your name, date, and a note on how much you can spare temporarily for someone else's emergency use. Store them upright, well above high water level, and sheltered if possible. Tilt the containers so that bungs or spout openings are at the highest position, to limit possible external moisture collection. (As you may know, sealed containers will, over time, produce water that has been entrained within the fuel.)

For extended flights, you can pre-position fuel at mid-point. If you're going for a pickup, you can haul cargo fuel, land, and refuel on the way out, leaving the containers for later pickup. But if you're taking a load on an extended flight out, you'll have to preposition earlier. Having pre-positioned fuel caches over a number of years in Alaskan wilds, the

author never had fuel taken by others without it being replaced within a few days.

Filtering Fuel

Always, repeat, always filter your fuel for water and other contaminants prior to putting it in your fuel tanks. And, of course, check the fuel at the tank drains and low points prior to each flight. In this regard, do the low points after you check the tank drains, in order to avoid the possibility of pulling tank contaminants into the supply lines if you check low points first.

When filtering fuel with chamois: use rough side up and gravity-feed only. Use only real chamois. The chamois will usually pass whatever first wets it; i.e., wet the chamois with gasoline first. However, if the chamois becomes saturated with water, it will start passing it! Therefore, avoid pouring highly contaminated fuel through chamois (don't drain to the bottom of the can if you are using old fuel, or if you know it to be contaminated).

Route, Terrain, and Weather

After reviewing weather reports/forecasts, plan your route so that, early in the flight, you can deal with the areas most likely to be affected by weather. (Weather reports and forecasts can age quickly.) For example, if you have the option of going through a mountain range along your route either early or late in the flight and weather could be a factor, all else being equal, go through early. Except

for occasional PIREPS while airborne, the weather information you have at take-off and what you observe enroute could be all you're going to get. It's amazing how quickly weather can change.

On the other hand, weather may force you to go through late in the above example. The point is to think of weather and route together while identifying weather-sensitive (and thus time-sensitive) points along your planned route.

Thoughtful planning of your route should include what you'll do if weather interferes at critical points along your route. The idea is to avoid putting yourself into "do or die" situations if fuel remaining and/or weather are not what you expected. Put simply, the bush pilot must always have one or more "escape hatches" handy and remain aware of their condition. If you don't, you are not in control of your own destiny – a situation to avoid by planning and alertness.

If your flight will be through a mountain pass your plan should be based on the <u>type</u> of pass it is. It's best to classify a pass in terms of <u>what you can do after you get into the pass</u>. If you can make a safe one-eighty-degree turn while in the pass, it's a "turn around pass." If you cannot turn around or land, then it's a "one-shot" pass and once you enter you're committed to go through, regardless.

Lengthy passes can be made up of several distinctly different types of passes separated by valleys. If you can safely land within the pass, then

it is reasonable to treat the segments separated by a landing area as individual passes. I call these "land-and-wait" passes. The "wait" part is for the weather to get better. If a long complex has no landing areas, then you should view the entire system based on its most severe restriction. Always plan for passes in terms of what you can do safely with your airplane, given terrain roughness, slope, and width, combined with the expected winds, ceilings, and restrictions to visibility. Never go through a pass for your first time in marginal weather conditions.

Some turn-around passes can be, and are frequently, turned into one-shot passes by high winds. Strong head or tail winds can make a pass one-shot in two different ways. For example, if you enter a turn-around pass that has tail winds close to or above your maximum cruise airspeed, you can forget about making a one-eighty. It's important to recall that surface wind speed can be tripled while flowing through a pass, depending upon pass shape. Strong head winds possibly can affect your making a safe one-eighty-degree turn. (This special situation and pass flying techniques are discussed in the following chapter.) If in doubt about the winds, assume the pass to be one-shot.

For any pass, but especially for one-shot passes, you need to know that the weather in the pass and at the far end is acceptable before you enter.

The IFR Option

There is always the IFR fly-over option for high terrain and passes. That's a good option if you, the airplane, and your procedures are legal and truly current. If your airplane is single engine, IFR is a leap of faith. That's not a big deal when you know and trust your engine. Here are some caveats.

First, the author will quickly note that some of his most memorable loads of structural ice have been acquired in turbulent clouds near mountainous terrain -- and they were not forecast or the so-called "known" icing.

Second, be sure your altimeter is good, even if it is legal, paper-wise. For instance, when weather presents the opportunity, fly by a peak with known elevation and verify your altimeter's accuracy.

Third, observe two thousand feet clearance above mountainous terrain within four NM of your track.

Fourth, check your pitot heat regularly. Loss of airspeed indication is not a calamity, of course, if you fly the desired pitch attitude with appropriate power and configuration.

Fifth, some pilots who legally fly IFR in uncontrolled airspace liken the probability of a mid-air collision in IMC as similar to that of a bunch of house flies buzzing around in a bottle and never

running into each other. But remember, the house flies are in VMC, have multiple-lens eyeballs, and can pull high-g turns on very short notice. Besides, there are surely some unreported wing-tip crunches inside that bottle. We human aviators simply must be picky about <u>which</u> uncontrolled "bottles" we use.

Finally, try your procedures in "severe clear" first. It will be valuable experience and might persuade you to plan a different route or to stick with VFR.

Using Offset Navigation

If flying to a particular place for the first time, consider using <u>offset</u> navigation on your last leg. That is, if your destination is not expected to stand out because of its characteristics, try intercepting a prominent nearby linear feature (shore, river, ridge, road, etc.) a few miles off to the right or left. Then, let this feature lead you to your landing area. This procedure takes the ambiguity out of "arriving," not seeing your destination, and then not really knowing which direction to search first. This old back-packer's orienteering trick should be a part of any pilot's skills. You never know, your GPS or Loran could go belly-up.

Your Navigation Plan and Your Charts

In planning your route, consider the following common situation. Let's say you're going to parallel a narrow mountain range that is being swept at a large angle by winds. If low weather is

not a factor, consider flying the windward side where you can take advantage of updrafts and a smoother ride as opposed to down drafts and turbulence in the lee. You'll not only be more comfortable but also save fuel. Additionally, you are more likely to see unexpected, incoming low weather earlier.

Prior to takeoff, have a simple plan of headings, distances/times on legs, and an estimate of fuel burn from take-off to the first available fuel. Keep your plan simple with only the basics; then you are more likely to make it and use it on a regular basis.

In addition to the usual aviation charts (Sectionals, WAC's), the author recommends USC&G 1:250,000 topographical charts. Because of their detail, "topo" charts are very useful in helping you discriminate amongst similar terrain features in mountains and in seemingly featureless tundra. The scale of these charts is just right for about one hundred twenty knots or less. They also have a lot of place names handy to know. Topographical charts can't replace the standard aviation charts; but they do make a very useful supplement for the low-level navigator. They're a bit pricey but you can cover them with plastic film and prolong their life.

Communications and Navigation Radios

During your flight in wilderness areas, monitor a communications frequency that will be of immediate use to you in case of sudden engine failure or other such sporting event. If the area is well traveled, that

frequency might be one of the multi-com frequencies that area pilots use. If not, then you can monitor the local Air Route Traffic Control Center frequency but, of course, not use it unless you are dealing with ATC or have an emergency. Local ARTCC area frequencies are listed in the Regional Airport/Facility Directories and the Alaska Supplement. You likely will be able to contact only high-flying airliners, but that's great.

An acquaintance once curled his prop at a remote spot, raised a Boeing 747 on the area's ARTCC frequency, the 747 crew relayed his message, and we had a replacement prop on the way in a short time.

In another attention-getting event, the author had an engine main bearing seize suddenly in a U-206 amphib at about 500 feet AGL, too far from useable water or suitable wheels-down turf. (Yes, there were ten-plus quarts of oil in the reservoir!) While gliding to an unscheduled stop on the smoothest available tundra, we "Mayday'd" on the local area multicom. Three airplanes were overhead within ten minutes. (The squeaky wheel gets the grease.)

Acquire and use a GPS navigation receiver. The Global Positioning Satellite system is the best thing that ever happened to navigation. Not only is it very accurate in pin-pointing your current position but also, most GPS receiver models can read out your ground speed and, indirectly, your drift angle plus other vital information such as latest ETA. Such information can alert you to unforecast wind

changes that could put you "in dire straits" fuel-wise. However, being at low altitude in mountainous terrain could deprive you of continuous GPS service. Also, there is the real danger that any pilot can become too dependent on GPS and lose visual pilotage skills.

Loran can also give you advantages in navigation, but is far more likely than GPS is to give false information and is not nearly as accurate, even at its best. Its accuracy depends on your distance from, and location relative to, the paired Loran ground-based transmitters. The future of Loran is uncertain.

In sum, GPS is the star of navigation and all else is "down in the grass." However, the bush pilot must maintain high skills in visual navigation and dead-reckoning in order to avoid sole reliance on <u>any</u> black box.

For example, during the early 1980's the author flew a busy year-round schedule inside of and along the slopes of the Alaska Range in the vicinity of Denali (Mt. McKinley) National Park and from southeast to west-southwest of Fairbanks. For the frequent situation of light falling snow and marginal flight visibility, it was very handy to use pre-determined radials from Fairbanks VOR in combination with flying along known sled trails first created by the wild meat market hunters of the late 1800's.

During the long twilights of Alaskan winters,

those sled trails really stand out against the black spruce forests while paralleling the Range and also when going to and from Fairbanks. After having the two VOR receivers in each of the heavier planes adjusted in a Fairbanks shop, we would run dual checks and known-position radial checks to ensure accuracy.

One time, however, I missed a turn point, got onto the wrong branch of the sled trails and came very close to a hill. The author was simply not doing a good job of dead reckoning and missed noting the failure of the VOR receiver in use. It was yet another cheap lesson.

The Destination

Among bush flying's biggest surprises will be that of arriving at a remote village airstrip and finding it being reconstructed by a grader, or circling what you absolutely know to be a previously used landing area and seeing it overgrown with alder brush.

Ordinarily, a prior radio or telephone call to the remote village can save a lot of trouble for you. Also, you can sometimes get the runway condition of remote strips from the nearest Flight Service Station. But with off-airport areas it's a different ball game.

Off-airport areas will change quickly due to human efforts as well as to natural forces such as precipitation/wind erosion, temperature

extremes, landslides, invader brush, plus, for beaches: ice, tides, currents, and storms.

It's hard to find an off-airport area where someone hasn't already been with an airplane. Sometimes, you can ask around and find someone familiar with your intended destination. Try to get a sketch of the area and the obstacles to flight operations, remembering that you both are talking about a place that could already have been drastically altered by the above-mentioned forces.

A sketch could even show that you're talking about two <u>different</u> places.

It is sometimes very useful to talk to knowledgeable passengers about your intended destination, remembering that even if the passenger is a rated pilot, he or she may not "have a clue" as to what your airplane's capabilities are. Local knowledge can be valuable, but it can also be hazardous. Consider, for example, the following "educational" experience of the author.

It was my first year flying for a float plane outfit called "Flirite" on Kodiak Island, Alaska. True to its name, Flirite, under the superb leadership of Chief Pilot Ralph Wright, was a professional, well-run air taxi service. We flew in tough weather but with great care and, for that reason, quite a number of the local private pilots and commercial fishermen flew with us as passengers whenever the weather became "wild."

On the day of this flight, Kodiak Island was under the influence of a vigorous low pressure and was getting "the works:" rain, gale force winds, and large bands of advection fog were lashing the eastern shores of the island. My task was to fly a commercial fisherman from the town of Kodiak to the village of Old Harbor on the southeast side. Old Harbor, shielded somewhat from easterly blows by the small island of Sitkalidak, was going "up and down" in ceiling and visibility.

At mid-morning our radio contact in Old Harbor called to say that the fog was lifting in Sitkalidak Strait to the northeast. Ralph gave the signal for me to give it a try.

We called our fisherman passenger. He was eager to rejoin his fishing boat. But, when he showed up <u>wearing</u> his full-body arctic waters exposure suit, he couldn't have stated more clearly his anxiety about flying that day.

The weather indeed made for some creeping, bouncing along in the U-206 floatplane, throttled back to 90 knots with half flaps, and maneuvering to stay in VMC along the eastern shorelines. Finally, we cleared Left Cape of Kiliuda Bay and entered wind-blown, fog-shrouded Sitkalidak Strait. The visibility was down to about two miles, I had my 1:250,000 "topo" chart out, and was tracking our progress carefully. I was "keyed up," but wasn't ready for what happened next.

Suddenly, the fisherman thrust his left arm

across my face, pointed to a misty bay entrance and announced, “We turn in there to follow the Strait!”

“No, sir,” I replied, “you’re looking at Aimee Bay, the Strait turns southward a little farther in. Turning now would put us into really deep trouble.” (Aimee Bay is ever-narrowing as you go in, with high steep hills on both sides -- a proven death trap.)

Accepting the challenge, he raised his voice even higher, “Now, listen, I know that you’re new to Kodiak! I’ve been fishing these waters for thirty-five years and I say you’re gonna get us killed if you don’t turn now!”

Pointing ahead, I said, “Look, over there is Shag Rock. Up there under the low clouds is the eastern shore of Sheep Island, where the Strait starts to turn toward Old Harbor.”

He repeated his demand. I started to plan my actions if he grabbed the controls in front of him.

Then, the low clouds lifted dramatically to reveal all of Sheep Island and, shortly thereafter, Old Harbor nestled along the Strait to the southwest. The cabin became quiet and remained so until after I beached the float plane. Then, softly, he said “Thanks,” got out with his gear and walked to his fishing boat.

Pilots, Charts, And Passengers

As illustrated by the above-described incident

and other situations, passengers do get concerned and want to help when a pilot gets out a chart. To them, when a pilot gets out a chart, especially in low weather, they believe he is lost or disoriented.

Probably, that's because motorists and some fishermen go to their charts only <u>after</u> they're lost.

Therefore, when you get out your charts, reassure your passengers that you're not lost and are using the charts so that you can remain oriented at all times. Having your charts out, in place, and ready to use before you start the engine helps a little in this regard, but not always. Being apprehensive is simply part of the make-up of many passengers.

Filing Your Flight Plan

<u>Always</u> file a flight plan—by telephone, SSB radio, VHF radio, or letter to the nearest Flight Service Station. Back it up with a note to your family or a friend, giving the basics of your plan. Include your possible alternative routes and stopping places, along with dates and approximate times.

Of course, it's best to file with the FSS. If you fail to do the smart thing and do <u>not</u> file <u>some</u> kind of flight plan, take plenty of camping, hunting and fishing gear, and food, in addition to your regular survival gear -- and <u>pray hard</u>.

CHAPTER FOUR - FLYING THE WEATHER AND TERRAIN

Your Alternative Plan: Keep A Back Door Open

Pilots learn early in training that flights don't always go as planned. So we're trained to handle the unexpected, like engine fire, unexpected weather, radio failure, etc. With good training, we also learn to avoid situations where there are no alternatives available if our plan is blocked. There is one theme in this book that is most important: to be a safer bush pilot you must always keep an alternative plan available. Put simply, you must:

ALWAYS KEEP A BACK DOOR OPEN

Situational Awareness

When flying in wilderness weather and terrain, "keeping a back door open" can be done only if you <u>know</u> your location and situation. That means staying aware of the fuel burn, your ETA, the <u>actual</u> weather compared to that which was forecast, the terrain ahead, your aircraft's current capabilities, the point-of-no-return due to fuel, the hours of daylight remaining, and what you'll do if your planned route is blocked by weather, etc. The list gets pretty long at times. So you learn to prioritize. That is, you watch closely the things that could quickly put your flight in trouble, while checking other items less frequently. (Hopefully, as a fledgling, you learned to check fuel, engine, and location every fifteen minutes.)

Remaining “situationally aware” is sometimes hard to do, as when having a load of fun-loving passengers or being treated to the splendid sights of the wilderness, but do it you must.

One of the best features you can have in a small aircraft's intercom is a pilot-isolate switch. With a headset on and “pilot-isolate” selected you can reduce distractions and focus on your situation. “Pilot-isolate” is good for rapidly changing situations such as low ceilings, low visibility, passes, takeoff, and landing. Keeping a simple fuel-time-position log also will help you remain aware of your situation.

Years ago, the author flew Navy fighters from aircraft carriers at night and sometimes in rather sporty weather. In the pilot's ready room there was a little sign that, in its own way, promoted situational awareness. It read: "If you're scheduled to fly tonight and you're smiling, you’re not completely aware of your situation.”

Nevertheless, smiling is okay, even encouraged, in the bush environment.

The Measure of a Successful Flight

Reaching your destination regardless of exposure to danger is <u>not</u> the standard of success. The first measure of success in civil aviation is to avoid unnecessary exposure to hazard. When first flying the bush, being recently a military pilot, the author had the idea that the objective was all-

important. And to him, the objective was the destination. The following experience helped correct that idea.

It was mid-April on Kodiak Island, the middle of the spring bear hunt and the air taxi outfit I'd joined two months before was very busy. My job was to fly from the town of Kodiak with a U-206 float plane to Red Lake, near the southwest end of Kodiak Island. There, I was to pick up two Alaskan resident hunters at the end of their ten-day bear hunt, It was about one hundred twenty nautical miles one-way; in lousy weather, over an hour of "hunkered-down" flying. The forecast was typical for spring: rain showers, heavy at times, southeast winds 25 knots, 35 to 40 knots in bays and passes, visibility three to five miles.

The weather was worse than forecast. But with the help of the air taxi pilot's guardian angel, and locked onto the notion that the destination was the "mission objective," I managed to reach Red Lake. Landing in the lee of a point where the hunters were camped, I then beached and secured the plane. The hunters watched me while sitting under a flapping, dripping tarp by their tent. They were nonchalantly sipping coffee and tending their campfire. “The rascals,” I thought, “they haven't even started to break camp on schedule!”

Sloshing to the campsite, I was proud to have made it to Red Lake despite the terrible weather. I announced to the hunters that I was there to take them back to town. The older of the two looked at

me with piercing eyes and said in a serious, measured way, "If you think we're gonna fly with any pilot dumb enough to fly in this (bleep) weather, you're full of (bleep)!"

I revised my flight plan by HF radio, accepted some coffee, and we waited out the weather. It calmed down three hours later and we left for town. The old hunter was right: the objectives of a flight must always include safety as the overriding concern, far more important than reaching the intended destination.

Day VFR Ceiling and Visibility Minimums

During the day, except for takeoff and landing, the air taxi airplane pilot must operate at or above 500 feet above the surface and 500 feet or more horizontally from any obstacle. For operations in Class G (uncontrolled) airspace, visibility must be at least two miles when the ceiling is below 1000 feet. (FAR 135.203 and 135.205)

In contrast, the private airplane pilot, operating in Class G airspace under FAR 91, legally needs only one mile visibility and clear-of-clouds at or below 1200 feet AGL but with standard VFR cloud clearances when above 1200 feet AGL. (FAR 91.155)

Reality check:

Often, the legal VFR ceiling and visibility minimums are not adequate at cruise speeds easily attained by general aviation airplanes. Chances are,

they will not be enough for high, rugged or sloping terrain. When you are not experienced with your route or don't have a linear terrain feature (beach, road, etc.) to follow for navigation, you should have a personal higher-than-minimums flight visibility.

Additionally, slow down! For example, following a fairly straight shoreline, a pilot might be comfortable in a small tail dragger cruising at eighty knots with one statute mile flight visibility. That pilot is doing a about ninety mph and is seeing about forty seconds' worth of air/shoreline ahead. (That's twenty seconds' worth for ninety mph airplanes coming at him head-on.) It's a tight setup. And it's a good reason to stay on the right side when you follow linear terrain features. Even a "puddle-jumper" needs to slow down in marginal conditions.

By slowing down, you increase your available reaction time for making decisions. That is, you slow the pace of events. Not only that. You also reduce the area you need for maneuvering or turning around. In short, by slowing down you reduce your own time-distance dynamics to fit the visible and usable volume of airspace in which you are flying.

Slowing down means getting on or near landing approach speed with ten to twenty degrees of flaps. Slowing down and <u>not</u> lowering flaps will restrict your visibility over the nose due to a higher angle of attack required to hold altitude.

If you're not comfortable with either the ceiling or visibility, it's time to turn around. Whenever you start wondering whether or not you should turn around, you're late -- do it now. That's a very reliable rule to follow. You always realize that later.

Minimum Cruising Altitudes

Over other than congested areas and, except for takeoff and landing and when over open water and sparsely populated areas, aircraft may not be operated below 500 feet above the surface. But even over open water and sparsely populated areas, you can't fly closer than 500 feet to any person, vessel, vehicle, or structure. (FAR 91.119)

Reality Check:

There are some good reasons for flying higher than 500 feet AGL. You're better off higher if your engine fails, you can see farther and more easily navigate at higher altitudes, your nav/com radios can reach greater distances when higher, you can react more effectively to quick changes in surface levels, and you can relax more, comparatively, at higher altitudes.

On the other hand, in the bush environment and due to changes in weather, you have to be proficient and safe in navigating at 500 feet AGL. Terrain does look different from low altitude with all the hills "chopped off" by cloud layer. (During my first three months on Kodiak Island, I would've sworn that all those hills had flat tops at about 700

feet.) Being proficient in pilotage at 500 feet AGL is achieved by flying at that altitude every time the situation permits. If you have justifiable confidence in your engine, you'll do well to practice navigating at legal, low altitudes.

Limiting Unnecessary Exposure To Hazards

Avoid unnecessary low flying over broad expanses of water or very rugged terrain where your chance of survival is severely limited if the engine fails. The idea is to reduce unnecessary exposure.

For example, the author has flown around wide bays of arctic water with a wheel plane many times rather than save a couple of minutes by cutting straight across. On the other hand, as part of the job, we had to fly across fifty-mile wide, icy Shelikof Strait and other open areas many times in small wheel planes. We usually did it as high as weather and performance allowed, although admittedly we couldn't get high enough to change the basic outcome with an engine failure. (It was a "feel good" thing at most.)

In order to avoid "white-out," avoid flying low over broad expanses of snow. Even though you can become comfortable in these situations, try to limit unnecessary exposure -- it's a matter of reducing statistical probabilities of undesirable outcomes.

As explained in Appendix A to this chapter, some of your extra altitude can be used as trade-off in

maintaining your airspeed during a tight turn-around. This is very useful during the turn-around in high wind conditions.

Even in light winds, so-called "nap-of-the-earth" flying at 200 feet AGL or less might be a good helicopter combat tactic but is not necessary nor recommended for ordinary bush flying. As explained in the chapter on wilderness resource survey flying, it is required for stream surveys of fish and occasional wildlife "quality" surveys. But as a technique for low weather navigation, it's unsafe and therefore unacceptable.

Surface Winds and Their Effects

The traditional way of visualizing surface winds and their effects is to consider wind as if it were flowing water. That makes it easy for the novice to understand that updrafts will be found on the upwind side of a ridge. And also that down drafts will be found on the downwind side. Therefore, if one is flying parallel to the ridge, then it's good technique to ride those updrafts on the windward side to save fuel and/or time, and to stay away from the down drafts on the leeward side.

The author's observations convinced him that the flowing water concept is useful but not fully adequate. Consider the following. Flying along the lee side of the Alaska Peninsula in order to avoid low, stormy weather on the windward side, several times the author has observed repetitive thick bands of disturbed water offshore and parallel to

the lee shore. There would be up to nine bands, starting at about one eighth-mile from the shoreline and diminishing in intensity towards the open Pacific.

Since he was driving a floatplane and traveling parallel to the shore, he investigated. As expected, there were down drafts on the shore side of each band and updrafts on the seaward side. The updrafts went up almost to 4000 feet MSL. Flying just outside of the innermost (and strongest) band of ruffled water put the airplane in a "bounce wave" updraft. Held level at the same power, it gained 5-8 knots for a long distance before Peninsula terrain rose high enough to disrupt the phenomenon. (See Figure 4-1, below.)

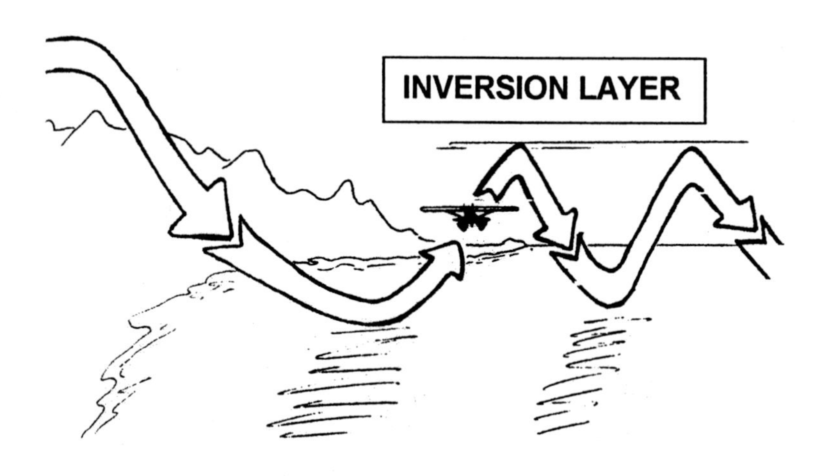

LEE SIDE BOUNCE WAVES
FIGURE 4-1

To the foregoing, some pilots might say "So what?" After all, strong bounce waves of wind are well-known in mountainous areas, especially by glider pilots. What was significant to me was the concept that, unlike water, air is compressible and therefore can retain its energy by compression, like a spring.

Therefore, wind and its effects are best considered as similar to flowing water, but compressible, and when hitting non-absorbent surfaces, are able to retain energy. You can put that energy to work on a practical level. The lee side down draft's energy could be absorbed by vegetation or rough terrain. Or, you might be able to find and ride a useful, energy-saving bounce-wave farther to the lee. Your fuel savings will be worth the effort.

Elevator Trim And Cruise Drag

On some aircraft, at cruise speeds you can change the elevator trim tab setting in order to "streamline" it and the elevator, while using elevator pressure to hold the aircraft level. The net result is that your airplane will increase its level cruise speed. The Cessna U-206 is one of these aircraft. Of course, holding constant nose-down elevator pressure to overcome extra nose-up trim will be tiring, but in low fuel or headwind situations it's worth it. Check your own airplane for similar fuel-saving characteristics.

Turbulence and Up/Down Drafts

When driving small airplanes in mountainous country, you can expect at least moderate turbulence where the winds within 5000 feet of the peaks and ridges are above 15 knots. When they're at 25 knots, you'll get a hard ride. Above 25 knots, those winds will produce mechanical turbulence and up/down drafts that'll make you wish you'd stayed home.

The 5-15-25 rule given above is a fairly good prognosticator. However, you're forced to plan your flight based on the winds-aloft forecast that can be based on hard data up to thirteen hours old. Good luck; plan for the worst.

In turbulent air, you'll want to have the airspeed above best-rate-of-climb speed (Vy), below design maneuvering (Va) speed (adjusted for weight), and in best flaps-up climb configuration. Being above Va in turbulence invites airframe structural damage or failure. Recall that design Va is correct for you <u>only</u> if you are at maximum allowable gross weight (MAGW), which is called simply "gross weight" by most of us.

With a lighter-than-max-weight airplane, the "operational Va" limit will be <u>less</u> than the design Va. (For the numbers-oriented: "operational Va" varies with the square root of actual weight squared over MAGW squared.). Design Va is in the pilot's operating handbook, but for a rough estimate of Va, you can multiply Vs1 (the bottom of the green band

on your airspeed indicator) by two, then subtract ten percent for a normal category small airplane. For example, if Vs1 is forty-five, then twice that is ninety, minus nine gives a rough estimate of design Va as eighty-one. This is on the safe side for an airplane with 3.8 g's as maximum positive load factor. Your "operational Va" will depend on how far below MAGW your actual weight is.

In rough air, set your airspeed above Vy and below operational Va. Cinch down your seat belt and shoulder harness, and lower your seat. When encountering a down draft, go to climb power and Vy, while doing a one-eighty if the down draft lasts more than a few seconds. Follow the axiom: "It's always going to get worse before it gets better." It usually does. Avoid rapid movements of the flight controls and try to keep the airspeed within the above bounds – all to avoid over-stressing wings and tail feathers.

If you encounter turbulence while also having to deal with low ceiling and/or visibility, let your intuition be your guide: turn around and try another route or go back. First, always avoid the deadly combination of low altitude or low visibility and turbulence. Second, always have much higher minimums for ceiling and visibility in rough air and/or rough terrain. These two rules are important to your longevity.

The companion of the down draft is the updraft that can put you into the clouds before you know it. Never go through a tight pass when close to a

ceiling. An unexpected updraft can whisk you into clouds and closely associated rocks before you can recover.

An updraft in clear air can also take you much higher than you want to be. A quick one-eighty, coupled with power reduction to idle and going to operational Va is the best you can do in a sustained, turbulent updraft. If the updraft is smooth and powerful, take your airspeed to the top of the green arc on your airspeed indicator.

When approaching high terrain head-on from the downwind side you might encounter the updraft first. Then, if you continue, you will experience the down draft. Depending on the terrain and your altitude, the first disturbance could very well be downward.

When paralleling high terrain on the lee side, the first big jolt could be up or down. Paralleling or approaching from the upwind side, you're more likely to experience only updrafts except near rapid terrain changes. You will learn quickly to turn away from high terrain with the first onset of serious up or down drafts. Even with manageable up and down drafts, you'll want to cross ridges with plenty of vertical clearance and at a forty-five degree angle. This makes it a quicker task to turn away from the high terrain.

When managing Denali Wilderness Air at McKinley Park in the early eighties, I remember giving the usual caution to a highly experienced but

newly hired flight-seeing pilot, about turning around promptly upon getting a sustained updraft in the lee of Mt. Mckinley. He had plenty of Alaska flight time but not much in the “big hills.” I could see he didn't really believe me when I explained the magnitude of air currents that can peg your VSI either direction despite what you do with pitch and power. Within two weeks, that pilot returned from a trip after such an experience. His eyes were wide and he tried hard not to hyper-ventilate when discussing it. It was all so powerfully smooth only one passenger, a rated pilot sitting up front, realized what was happening. Safe mountain flying got another convert that day.

Avoiding Destructive Winds

Ordinarily, air turbulence and up/down drafts are caused by pressure gradient winds (the high-pressure-to-low-pressure kind) flowing over rough terrain and mountains. Pressure gradient winds also cause or augment down-slope winds on the lee side of mountain ranges. During their plunge, these winds accelerate. The result can be moderate to extreme turbulence downwind of a mountain range when the winds aloft within 5000 feet of the ridges exceed 15 knots. This thumb rule will vary for different locales, but you'll do well to be alert to the possibility of these katabatic winds, called "Chinook, Taku, Santa Ana," etc. If you are approaching the downwind side of a mountain range you might see lenticular clouds on your side above the peaks and ridges, but only if there’s enough moisture aloft. In dry air those tell-tale

clouds might not be there.

Also, look for dust being kicked up by winds in dry creek beds and treeless open spaces. If you see just a hint of dust in the air near the surface, expect katabatic winds with impressive turbulence near the surface and stay up and away accordingly.

Other types of katabatic winds include the effects of heavy cold air sliding down mountain slopes from high snow/ice fields and glaciers to pound the valleys and creeks below. They also will kick dust into the air at the feet of glaciers. Their speeds and destructive capabilities are intensified by the sun's heating of vegetation and terrain at the lower elevations without equivalent heating in the reflective ice and snowfields of higher levels. These effects are most intense in the spring and early summer, mid-morning or so, on sunny days.

When doing springtime caribou surveys in the Alaska Range, we got started early. We would depart the airstrip with the biologist at earliest light (about 0300) to avoid sticky snow takeoffs and, even more important, to avoid the katabatic winds of late morning. These winds would turn valleys and passes into boiling cauldrons and make for wild rides if we finished late. (Nature's negative reinforcement does wonders for late starters.)

The passes were especially wild in the winds of late morning and we did everything feasible to avoid flying passes, especially the narrow ones.

Glaciers

Take care near glaciers and their lower reaches especially during spring and early summer, due the katabatic winds, discussed in the preceding section on "destructive winds." Most glaciers could be in areas (parks, etc.) where you need permission to fly below specified altitudes as well as permission to land an airplane. Get any necessary permission before you fly into such areas.

Here are three safety rules about glaciers. First, never intentionally fly below the level of the side ridges confining a glacier unless the wind is light to calm and you have a good reason to be there. Second, if you must fly below the ridges due to weather, never fly up a glacier, due to the rapid surface changes. Third, if you intend to land on or fly low over a glacier, take an experienced glacier pilot as pilot-in-command on your first flight.

Glaciers have large, near-invisible surface changes, depending on the light, and frequent violent wind shears. We're not discussing the more stable, broad, snow covered accumulation fields at the head of many glaciers, to which most commercial operators fly skiers and climbers. But, they too can harbor violent winds. We are discussing the more normal rougher, steeper, reaches of glaciers below the upper accumulation field. Stay away from areas below that upper field.

A good spot will, with light winds, permit landing up slope and taking off down slope in order to help

compensate for reduced power at higher density altitudes. This must be done with care while inserting, for example, mountain sheep hunters, photographers and/or technical climbers. Good weather and light winds are prerequisites.

Another prerequisite is to drop small tree boughs or weighted, dark, plastic bags along your intended landing and takeoff areas to improve depth perception and help avoid "white-out."

Never land on a glacier in an area that you haven't recently visited or previously photographed and studied. By taking the photos in low-angle sunlight, you can spot rough surfaces more easily. Nevertheless, you still can't see some bad spots until getting near touchdown. It takes a lot of care as well as luck to avoid making your airplane a permanent part of any glacier. Despite your precautions, there are always the crevasses under snow and thin ice. Although slow in moving, glaciers are ever changing.

You can count on the crevasses and holes being there and should always use safety lines clipped or tied to landing struts while working in the vicinity of the airplane. If going farther from the airplane it's a good idea to tie-in with other people as well as using ice axes, with readiness to self-arrest.

You can expect quick changes in weather, especially the winds, so it's a good idea to stay close to the airplane and be ready for quick departure. Over-night stays are not recommended.

Furthermore, the higher the elevation, the quicker the weather changes, and the poorer is your normally aspirated airplane's performance due to the higher density altitudes. Thus, the lack of turbo-charging or normalizing can invite serious problems on takeoff. This makes a down slope run even more important for your takeoff. The best advice regarding glaciers is to stay well above or away from them.

Low Level Mountain Gales

It is not uncommon for mountainous areas to have "surprise" low-level winds above 35-45 knots with no frontal system associated with the winds. While living and flying within the Alaska Range of interior Alaska, the author would see this happen when a high pressure area moved nearby with a tight inversion layer setting up. Then the low level gales (and the fun) would begin.

If the destination's surface wind was tolerable, we could get there by careful planning. Our home strip was long but surrounded by a forest of tall White Spruce. When the gales blew, they were invariably at an angle to the dirt strip, and so, below treetop level, the wind sheared to almost calm.

After lift-off the author would hold the aircraft well below tree top level in order to get above Vy prior to climbing through the wind shear. Then, we'd climb on the downwind side of our valley until getting above the inversion layer, near the level of most peaks and ridges of the Range.

Above the inversion layer, the winds would be very light. We always made the letdown at between Vy and Va, on the downwind side of any valley, just in case the reported destination winds had increased and turbulence was present. Upon return for landing, we had to be careful to carry an extra 15 or so knots, with flaps up or only ten degrees, depending on the airplane, to be ready for the certain, quick wind shear to calm. When the wind sheared, flaps were extended as the airspeed decreased and a normal landing completed.

The fact that our strip was protected on the surface by tall trees permitted us to taxi safely in all high winds that were not aligned with the strip. This is a good demonstration that the limiting factor in windy conditions could simply be surface handling rather than flying conditions.

Depending on the season, mountain gales at low levels will show their presence with blowing dust and snow in the valleys. Knowing about this special kind of winds could prove useful to you some day.

The Adiabatic Process, Clouds, and Stability

The wilderness pilot must be informed on basic weather processes. That is fundamental to recognizing what is going on around you in flight, and therefore what can be expected. In short, you have to become a skilled amateur meteorologist, which means you have to keep working at increasing your knowledge of the subject.

A good source to start with is the book, "Aviation Weather," Advisory Circular 00-6A, for sale by the U. S. Government Printing Office. It's clear and basic. Study especially chapters 5, 6, and 7. There are other excellent aviation weather books on the market. Purchase and read all you can; it really pays off when you're in the cockpit and trying to figure out what's going on with the weather.

Understanding especially the adiabatic process, clouds, and stability will open the door to understanding weather you see in flight. As a consequence you will make better weather-related decisions. Here's but a small example.

More than once, in strong southwesterly winds, we've flown into the Healy end of Windward Pass in the Alaska Range, found a layer of stratus over on the downwind side and got under it for a smooth ride through the pass. At the same time, the rear seat biologist and I would see other airplanes over in the sunlight, under broken clouds (read "large adiabatic holes caused by compression-heated down drafts"). Later, we would listen to those same pilots reporting severe turbulence to Fairbanks FSS. In response, we would do our diplomatic best to spread the word.

Fog

If you takeoff at first daylight, as a matter of practice, be alert. When there is much moisture in the air, you could yet see the formation of fog prior to sunrise, simply because the air at ground level

could still be cooling due to radiation loss of heat from the underlying ground. The most likely time for radiation fog is at sunrise. Unless higher clouds move in or the sun's elevation remains low, radiation fog usually will "burn off" and allow you to fly.

If you fly in an area near large bodies of water, expect advection fog at some time or another to move onshore. Advection fog can be dense, fast moving, and very deep. It is therefore not easily "burned off" by sunlight on land. It also can move in starting as a wisp and sometimes give you opportunity to take the right decision before it really socks in.

On maritime shores, fog will usually (but not always) "telegraph its intentions" to come ashore by collecting on points and capes first. When you see that happening, count on fog spreading along shores and onto land. Another rule: if the fog lifts, wait at least a half-hour before you go flying. You'll be surprised how many times the weather comes down again before lifting to stay, or it might stay down.

Never fly in or near fog. Unless the fog is a clearly defined fog bank extending to the surface, it is near impossible to know how far you are from it. Here's an experience not to be proud of, but it made this pilot a believer.

It was during salmon fishing season on the Bering Sea side of the Alaska Peninsula. The

author was flying for Alaska Department of Fish and Game and was also a Peace Officer for enforcement of Fish and Game Code. I had flown to a fish camp on Nelson Lagoon to investigate complaints from a fisherman, and would relay that information to the area's Senior Management Biologist.

As we finished our meeting, Bering Sea fog started moving across the lagoon toward the beach where my Super Cub on big wheels was parked. The fog was coming in with the tide. At first it was thick and it looked like I'd spend the night there. Fifteen minutes later, it lifted to what appeared to be at least 500 feet, with visibility over one mile. Eager to get out of there, I took off and immediately realized that I was simply in an area between two thick windrows of fog. With flaps down and slowed to approach speed, I started circling to remain between the two rows of fog. I should have returned to the beach and pushed the airplane to higher ground.

Instead, I continued circling to stay between the fog windrows as they moved southeast toward Herendeen Bay. Surely the land's warmth, I thought, would thin out the fog even more as it drifted inland.

Then, I could follow the beach to the upper bay that was usually shielded from Bering Sea weather by adjacent mountains. My plan didn't work very well. It was a sucker's trap, and I was the sucker.

As the fog drifted towards Herendeen, it thickened and filled in between windrows. I was

forced down to a sporty job of circling while avoiding the high, grass-covered dunes just as I reached the safety of the bay shoreline. The remainder of the plan worked. But, never again!

Dealing With Freezing Rain

Whenever you're flying in clear air that is at or below freezing, and you are under clouds, be ready for the possibility of super-cooled rain coating your airplane with ice. If your area is coming under the influence of a warm or occluded front, the weather service could forecast freezing rain for this situation, and you can be forewarned. If the rain is developing as widely scattered showers with freezing rain, the situation usually becomes very uncertain and hazardous. So, don't fly.

However, in mountainous areas, and this is just an opinion based on several surprise encounters, there are opportunities to get unforecast freezing rain. It's simply the mountains squeezing more moisture out of the air than was expected. The wilderness pilot has to look out for "number one." So, you learn to obtain the forecast temperatures (low and high), check the area forecast for clouds at warm levels and then act, depending on the way the weather develops, including making sure the windshield defrosters are working well.

The amphibious DeHavilland Beaver flown by the author did not have a defroster but heated air from the circular engine cowling drawn directly onto the windshield was of some small help. Therefore, the

Beaver was more restricted than the other planes in these situations.

Anytime you have the possibility of freezing rain, you should, first and foremost, keep the windshield warmed for immediate melting of the ice. The defroster won't get all of the ice, after all, it's designed to handle the inside of the windshield. But it will be much better than not having had the windscreen already warm.

Second, if freezing rain coats your plane, quickly exit the area you just entered. My technique is to do a quick check of six o'clock and both sides for room to turn, check the D/G for the reciprocal course. Then, I make a one-eighty turn (part visual, part instruments).

Third, find a way to get around the shower visually. That takes luck.

Finally, with a float plane over salt water (downwind from the mountains producing the freezing rain), descend to a safe but low altitude over the water to get the benefit of the warmer air that might be there. Caution: if the air temperature is about 28 degrees or less, don't go low over salt water when the wind is up, you could easily pick up a load of sea spray ice.

Following Creeks, Rivers, and Roads

The use of creeks, rivers, valleys, canyons, and roads as linear features for navigation is usually

good practice as long as you make certain identification of the feature. It stands to reason that you're more likely to turn into the wrong drainage when you're going upstream. Going downstream, creeks and rivers run together and get bigger and fewer, making navigation simpler. Stay to the right side but give plenty of room for the airplanes coming at you that don't.

Take care when following snow-covered, paved highways, especially at junctions. You might mistakenly shift to a dirt road that will lead you into a hazardous terrain situation. It has happened; paved and unpaved roads look alike under thick snow.

Identify the highest elevation on the route that your navigational feature will reach. Then, if it is higher than where you pick it up - that is, if you're going upstream or climbing - add 500 feet (minimum) to that elevation, and climb to the altitude you'll eventually need. You'll usually find out, earlier, whether or not there is a possibility of your having to turn around due to low ceiling. This is not infallible. Low ceilings are sometimes lifted by changes in wind as elevation increases or by going into drier air.

When following a low feature such as a creek, river, valley, or canyon, stay in a position favorable for turning around. Often, especially if you're slow, you need altitude to trade-off in order to maintain a safe airspeed during turn around. (Appendix A gives an analysis of the turn-around maneuver.)

Even in remote wilderness, expect that there <u>will</u> be uncharted cables and lines across creeks, rivers, and canyons. They're waiting for the pilot who mistakenly thinks they'll have those civilized orange balls on them. Always stay well above and to the side of the feature and keep your "back door open" by keeping adequate altitude and visibility for the turn-around.

When below 1200 AGL, the author stays about 100 feet below the lowest "bumps" of the ceiling, or whatever it takes above 500 feet AGL to maximize visibility ahead. That gives the best view and also lets one keep track of what the ceiling is doing vertically simply by looking at the altimeter. The altimeter may be off a little, but that doesn't matter – you're interested in the <u>trend</u> of the ceiling.

Second Ridge Technique

As you may know, when you closely parallel a mountain range, you'll frequently be crossing a series of shoulders or sloping ridges at angles. These ridges or shoulders confine drainages (creeks, rivers) that can provide convenient "escape hatches" in marginal weather. When the weather is marginal, you might try what the author calls the "second ridge" technique. You fly as high up on the ridges as the ceiling allows, at a contour level giving you the best visibility, seeing on the topographical chart just where and in what direction from you that your contour level "cuts" the next succeeding ridge.

You can estimate the direction and distance to each "second ridge" at your contour level (altitude) as you cross one before it. If you don't start seeing that ridge by the time you think you ought to, then you may have to turn outward and start descending to find out what's happening to the weather. If the ridges are fairly close together, you'll likely be able to see the third or more ridge from your crossing point. This technique requires constant situational awareness – just what you want in marginal weather -- and requires you to stay VFR.

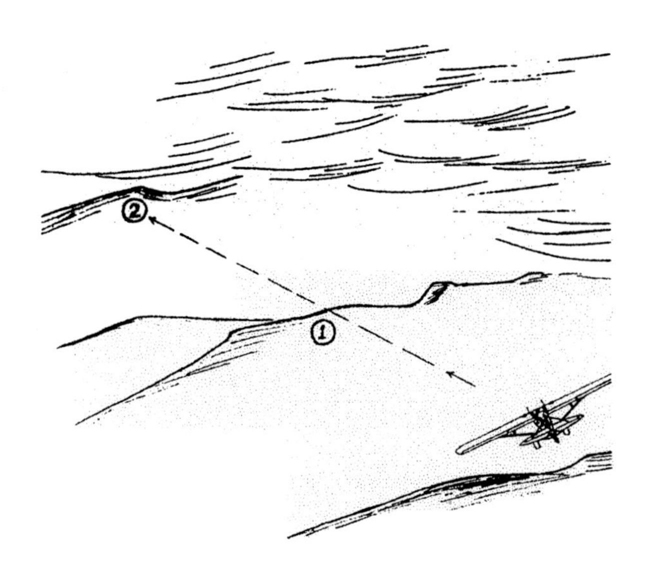

SECOND RIDGE TECHNIQUE
FIGURE 4-2

Spring Breakup Problems

An aggravating condition at high latitudes and at some higher elevations, that occurs every year, is spring breakup. ("Breakup" refers to ice in creeks

and rivers.) For the uninitiated, this is also when the "whole world turns into mud." The problem is that the surface is thawed only to several inches down and the still-frozen ground below can't absorb the water from the mud above. This condition is relieved by an extended, steady, and relatively warm rain. The warm rain thaws the deeper ground sufficient for it to absorb surface moisture. When that starts, it's best to get your strip free of ruts before it dries solidly.

In areas where surface or near-surface soil is permanently frozen (permafrost), breakup may or may not be a problem. It depends on how far below the surface the permafrost lies, the available drainage, and how much rainfall the area receives. Some of these areas remain completely unusable during the warmer seasons. The important thing is to be looking for this problem and to call ahead.

Using a nearby beach or graded road is the best solution; after any required approval from authorities and posting a safety lookout with radio.

Flying Mountain Passes

Never try to enter a mountain pass without knowing all you can about that pass, including its ceiling, visibility, wind, and turbulence, as well as what the weather is at the other end. It is important to do your best to fly any particular pass for the first time in good VFR weather with light or calm wind and little turbulence. The author likes to fly a pass at about 1000 feet above the confining peaks and

ridges the first time and then, down at 500 feet AGL and at slow cruise on the second "go." In any case, it's best to have extra high weather minimums when flying a pass for the first time.

Knowing all you can know about a pass also includes knowledge of the terrain near both ends. Most often, the approaches will slope upwards towards the pass. This can set you up to be <u>low and slow</u> as you enter -- <u>not</u> a safe situation. Avoid this, as mentioned before, by getting on pass elevation plus at least 500 feet well before you reach the entry point. (The 1:250,000 topographic charts are good sources for most pass elevations.)

If there are stronger than light winds, you can expect wind shears and an increase in the prevailing pressure gradient winds after entering the pass. You may be forced to maneuver as well as to counteract up and down drafts during the initial phase, especially if the winds are above ten knots or so.

With adequate visibility, staying between Vy and Va (for the airplane's current weight) is a good tactic. In a “turn-around” pass, you should try to maintain lateral position so as to be ready for a turn-around for any reason. As explained in Appendix A to this chapter, it is essential during turn-arounds to have good altitude, airspeed, and bank discipline. If the pass is wide, there could be a downwind side, which usually offers a smoother ride, possibly better performance, and the desirable turn into the wind when initiating a turn-around.

Surface winds being funneled through a pass can be tripled in speed. Due to the venturi effect, your altimeter may start to read high. (Venturis will speed-up the flowing air and reduce its atmospheric pressure.) Your ground speed, of course, could be changed drastically. Watch for that.

There are many valley-to-valley short passes that permit you to see the far side of the valley on the other side of the pass. These little "see-through" passes often will show you unique patterns on the far-side hills -- and the patterns will vary between summer and winter due to different accumulated snow patterns, along with vegetation changes. Try to memorize their visual "signatures." They're very handy when light snow or rain is falling, visibility is down, and you want to go through a short pass to the next valley without penetrating cloud on the other side. If you don't see the "signature" on the next valley's far side, don't enter the pass. That makes decisions easy to take and avoids uncertainties.

In these same conditions, avoid passes that accumulate snow evenly on both sides, and thereby will disappear to put you in "whiteout" with only a slight decrease in visibility from falling snow. The best kind of passes are those with vertical rocks that accumulate very little snow.

When picking the pass to fly, choose only the sure thing. Be skeptical.

The "Historical Precedent Effect"

A number of years ago in Fairbanks, I saw a young pilot that I had known before. When we first were acquainted, I was supporting a big game guide and he was an assistant to the same guide in the Alaska Range. Now, he was a licensed Commercial Pilot and about to embark on flying support for another guide in the interior of Alaska. Before we parted, he asked me if I had any advice. I told him that he should set his personal weather minimums at a safe level and never break his own rules. If he does break his own rules, he should make sure the guide understands that it was a mistake and he would not, for safety's sake, do that again.

I saw him again a year later. He had “pranged” a Super Cub the previous autumn while flying in foggy conditions in the Brooks Range. It seems he had broken his own minimums successfully once and for the remainder of the season was under pressure to duplicate his luck. So he tried, with the inevitable result.

Private pilots are just as subject to the "historical precedent effect" as commercial pilots. It's not easy, but do make yourself immune to it by never violating your own minimums – for any reason. The more times you say “no” under pressure, the easier it is to say “no” in the future.

APPENDIX A TO CHAPTER FOUR
THE “FATAL DOWNWIND TURN" -- THE AUTHOR'S INVESTIGATION

Over the years, the author has read a number of articles on the "fatal turn downwind" in aviation periodicals. Essential elements during the maneuver were low altitude and high winds. Many pilots will swear that making a turn to the downwind in a strong wind is the same as experiencing a wind shearing to the tail. Therefore, the argument goes, you are flirting with not being able to accelerate quickly enough to avoid a stall. Physicists and engineers will tend to deny that notion on the grounds of conservation-of-energy principles.

In the early eighties, I saw a chart of Alaska showing the locations of small airplane crashes. A large number of crashes were in or near mountain passes. It's reasonable to speculate that a number of those crashes occurred during turn-around.

With my jobs flying busy schedules from within the interior Alaska Range, as well as in the rugged terrain of Kodiak Island, lower Alaska Peninsula, and eastern Aleutian Islands, I took frequent opportunities to investigate the turn downwind. I tried turns downwind and upwind, at 500 feet AGL and at 200 feet AGL, in passes and in open areas, and in high winds and with no wind. The investigation took a few seasons before I had confidence in my conclusions.

The Investigation

The first objective was to understand the problem, if possible. Because I did a lot of caribou herd quality surveys, and stream surveys (for fish) at 175 to 250 feet AGL, I was comfortable with the altitude, but there was one difference: the winds.

When doing fish surveys, we would avoid winds above twenty knots that make it difficult for the biologist to see through the choppy surface of the water to count and classify fish. Or in land animal surveys, the wind and turbulence would be too strong to permit safe low altitude surveys. Moreover, land animals will tend to “hole up” in their favorite hiding spots during windy weather. Herd animals become “spooky” making quality surveys difficult.

Therefore, I did the almost all of the investigation when on solo freight hauls to wilderness stations and on solo salmon stream-marking flights to ocean beaches. The flights were in the DeHavilland amphibious Beaver, Cessna 206's on floats, wheel-skis, and wheels, Cessna 180/185’s on wheels and wheel-skis, and in big wheel, floats, and ski-plane Piper Super Cubs. Almost all were at or near maximum allowable gross weight (MAGW).

The turns were done at three different angles of bank: 30, 45, and 60-70 degrees, as accurately and as steady as I could make them. (The Cubs had only a turn coordinator or turn-and-slip indicator.)

The Results

When turning at bank angles near or above 45 degrees, all of the planes would lose airspeed noticeably if power were not added or the nose not lowered to trade-off altitude for airspeed. Of course, the speed loss was quicker when starting at the slower speeds that are appropriate for low altitude and/or low visibility navigation. As expected, when at 200 feet AGL, it was not safe to exceed 45 degrees bank in a turn at MAGW when starting at slow airspeeds. Even starting at cruise speeds between Vy and Va, the large amount of induced drag created in a 45 to 70 degrees banked turn is evidenced by rapid speed decay. This was more pronounced in the Beaver and Cub. (So far, no surprises.)

An important factor is what the pilot perceives during low altitude steep turns in high winds. In a turn upwind, the pilot gets the visual sensation that the airplane is not turning enough for the bank established. Based on my experience as a CFI, the response of some inexperienced pilots likely could be to add inside rudder or increase bank, or both. This raises the possibility of a skidding turn stall as the airspeed decays. Recall that the low wing snaps even lower in a skidding turn stall. Skidding or not, increasing the bank will increase induced drag, stall speed, and airspeed decay without substantial power increase or loss of altitude.

The turn downwind is exciting when done at low altitude in close quarters with high wind. One gets

the very strong impression that the airplane is turning too quickly for the angle of bank. Every time I tried the downwind turn in a pass with strong winds I got the urge to "bring the airplane under better control" by reducing the angle of bank. That sensation was present even with only 30 degrees of bank. Of course, reducing bank is not the thing to do in tight quarters if one wants to avoid colliding with rock.

In both kinds of turn, upwind and downwind, the strong chop in the air adds a little spice and the airspeed needle jumps all over the indicator. In the real situation, add some low weather, which is likely the reason you would be turning, and your visual perception of what's going on could become saturated. The closer you are to the ground, the greater are the false perceptions.

One would have to use a ground based, three-dimensional tracking and recording radar coupled with a recording anemometer to analyze what is actually going on. Therefore my investigation should not be considered as scientific, but I did reach some subjective conclusions.

The Conclusions

First, for the most part, my experience in this non-scientific investigation suggests that the physicists and engineers are correct to a degree when they say that it's not the physics of the downwind turn that creates danger. Indeed, what I experienced was largely consistent with the

aerodynamics of turns and the visual perceptions created by large wind drift angles while turning at low altitude.

Second, despite the first conclusion, I got the clear impression that it was more difficult to maintain one's airspeed on the turn downwind than on the turn upwind. I am convinced that this difference exists. That supports the side of the argument emphasizing the effects of momentum-acceleration created by the turn downwind, thus likening such a turn to a head wind shearing to a tail wind.

Third, the danger comes from a combination of (1) the aerodynamics of low speed, steep-banked turns and (2) the false visual perceptions of the pilot, especially in downwind turns with strong winds. It takes a good bit of self-discipline to ignore the misperceptions while being precise with stick-and-rudder. Furthermore, the misperceptions are magnified at low altitude.

Fourth, even if it's never settled scientifically, settlement is not as important as having good procedures for making a safe turn-around in high wind conditions.

Fifth, pilots should avoid low ceilings or low visibility in and near rough terrain, or in high winds. These are deadly combinations that can take their toll of errant pilots in low level and turning flight.

Recommendations

The author's recommended turn-around is one that is planned for about 60-70 degrees of bank when above Vy (best rate-of-climb airspeed), and for 45 degrees when below and then, reducing to 30 degrees of bank as you slow further. You should lower flaps to 20 degrees when below Vfe. Trading off altitude for maintaining airspeed should be done only when full power doesn't do the job of maintaining safe airspeed.

You should expect and try to ignore the weird perceptions of a turn at low altitude in high winds, especially in a turn downwind. Bank angles appropriate to the indicated airspeed must be maintained when in tight quarters.

It is strongly recommended that you practice your turn-around procedure (at MAGW) at safe altitude until you have it well in hand. Then, within the restrictions of FAR Part 91, try it at successively lower altitudes, down to 500 feet AGL with lots of room in which to maneuver. Be smooth on the controls and do your initial practice turn-arounds in smooth air, in order to get a good "feel" for the airplane.

The goal is to develop solid, safe, airspeed and bank-angle discipline during the turn-around maneuver. That takes practice.

CHAPTER FIVE

BEFORE LANDING: PLAN THE TAKEOFF AND LANDING

You Need A System For Pre-landing Evaluation

If your destination is an off-airport site or a little-used wilderness village strip, you've come to the right chapter. For <u>any</u> uncontrolled airport, no matter how often used, it is good procedure to do a minimum of one turn or a fly-by over the runway at or above traffic pattern altitude. Do this, if for no other reason than to look for wildlife including, for village strips, the kids on three-wheelers.

For those spots that are not used often, you need to get down closer to see things that will directly affect your safety. The wilder the proposed landing site is, the more detailed must be your scrutiny. If you have a <u>system</u> for conducting your evaluation, you are less likely to overlook hazards.

This book gives a system that compensates, in part, for the lack of site preparation and maintenance. That lack makes the difference between on- and off-airport operations -- and consequently the big difference in safety.

Off-airport operations are inherently dangerous. The objective of airborne evaluation is to <u>minimize</u> as much as practicable the hazards.

The hazards can be further reduced by using

your resourcefulness and changing your approach or departure procedures to fit the situation when simple changes will do. If you need some fancy flying to avoid hazards, reject the site. Just stick with plain vanilla flying, but do it with solid precision.

Be ready to say "no" to an off-airport site if warranted. No matter how often you've been there, or how recent your last visit, the site can and does change. That's why you need a system -- a mental check list -- for your evaluation of any and all uncontrolled or off-airport sites prior to landing. Once you develop the habit of using your system, it will stand you in good stead. Furthermore, if you make it a habit, you'll always feel the need when you try to skip it. That's good.

Good Vision Required

In chapter one, we stated that bush flying is an intensely visual endeavor. Evaluating a prospective site for your landing and takeoff is a perfect example of the visual requirement. Takeoff and landing accidents are not always blamed on faulty evaluation. However, thorough evaluation prior to using a site can prevent accidents. Having good vision can result routinely in important decisions like rejection of the site, a small but critical change in your touchdown point, making a turn on water before coming off the step, a non-standard departure after takeoff, etc. In short, if you need glasses for normal distant vision, wear them.

As a major example, the author recalls arriving at an accident scene shortly after the two-seater, just after liftoff, had flown into the <u>only</u> tree on that very long, wide gravel bar. Both occupants were bruised somewhat but not seriously. While flying them to the nearest telephone, the author asked the pilot what happened. He said, "I forgot to put on my glasses."

The Evaluation

Be very methodical in your evaluation and remain skeptical of everything you see. Explain to your passengers what you're doing and why it's necessary. As you evaluate and plan your takeoff and your landing, if you keep wondering what's down there that you haven't seen, you're in the right frame of mind. If you keep seeing new important things, keep looking. Don't rush and don't let anyone else rush you. And don't lower your minimum standards for anyone. Later, you'll be glad you didn't. So will they, especially since they're aboard the airplane. Here's an example from the author's experience.

A government agency had two people in the field and planned to pick them up on a lake. It was August; the time when area lakes fed by snow in nearby hills are at their lowest when without recent rains. Several days before the scheduled pickup, the agency's floatplane was trashed in a high winds accident. They requested that we make the pickup

at a lake that had looked shallow that summer. When I questioned the advisability of taking our amphibious Beaver into the lake due to what looked like shallow water, they insisted that a twin engine Grumman Goose had been used there several times before, at the same time of year. (The pressure's really on when they use historical precedent.)

Still skeptical, I put on hip boots and flew our big-wheel Super Cub out to the lake, landing on its short little beach. Then I walked all over the lake -- in water below my knees. The Beaver could have spent the Alaskan winter in that lake, on its back.

A few days later, after waiting for strong winds from the right direction for the tiny beach, I retrieved the team with a few reduced Super Cub loads. An experience like that strengthens one's resolve to resist onslaughts of "urgent desire" against "cautious reason," a common situation in the bush.

Surface Wind

Start your off-airport evaluation by checking the surface wind strength and direction. This is usually not a big question unless the wind is light and variable. If so, look at the surface of nearby ponds or lakes to find the wind shadow on the upwind side. (For the non-seaplane pilot: the "wind shadow" is the strip of calm water along the upwind shoreline.)

Look at the direction in which birds face when standing on the ground, and in which they takeoff.

Take care, you will see a few unskilled birds taking off in goofy directions, especially when the wind is light. Birds that soar are lightly wing-loaded and sometimes takeoff downwind.

The foliage on the upwind side of trees and bushes often show the lighter green of leaf bottoms.

Flying at approach speed across the landing area at a big angle to landing direction will help you determine the direction of light wind by checking for drift. For two-way areas, if the length is marginally acceptable, you can even fly a good portion of the glide slope (towards specific aim points), both ways, and find which direction requires the most power to hold the same glide angle at the same airspeed. The direction that requires the most power is upwind.

Flying level at a constant airspeed, you can fly both directions along the landing area, timing its length both ways. The longest time is, of course, into the wind. (Be careful; there could be a lower wind shear present to ruin this measure.)

Plan Your Departure

Visualize your departure and what the wind could do to you after liftoff. If take off and climb are made unsafe by wind direction or terrain, you're looking at a one-way operating area.

For a one-way area, if you are contemplating landing a floatplane on a river, first consider the expected takeoff wind/current situation. Or, consider two different areas: one for landing and another for takeoff, with a clear taxiway between.

Be sure you evaluate the takeoff area before you land. In many rivers, you may have to get on the step, reduce power to follow a river bend on the step, and then complete the takeoff when the river straightens. Now, integrate the expected wind effects on your directional control as well as your radius of turn. For that, or with a narrow river, consider having water rudder(s) down for take-off.

Often, it will be advisable to get on the step and then throttle back to maintain a barely-on-the-step speed so as to minimize your on-step turn radius. In this situation, any turn into the wind must be done with special care to keep the wings level, with readiness to abort at any time.

For landing in a narrow river where you'll need precise steering, you might require water rudders down. But first, make sure you have adequate bumper(s) on the stern(s) to prevent the rudders from damaging float transom(s) when they flip up on touchdown. In most situations, a quick lowering of rudders immediately after touchdown is better. The location of your rudder up-down control could be the determining factor. You'll want to keep head up and eyes out of cockpit.

Use ailerons with the rudder, of course, when turning on-step or when needed to stay level. Be very sensitive to wind direction and how it might upset you on any turns, especially for turns into the wind. Allow for the current in all your maneuvers. There will be occasions where landing into the current is required. Do not touch down fast and flat or the plane will try to tuck under or even diverge laterally – not good! Also, expect a "hard" arrival.

A strong head current certainly can get your attention if you can't "go with the flow," so take care to be slow at touch down.

Important rule: never land or takeoff in water where you can't see the bottom unless you are certain of its depth and its freedom from rock as well as submerged or floating debris.

Study the bottom carefully, looking for the deep channels and "lining them up" with reference points along the banks. Identify the rocks and shallow areas. Then plan your takeoff and landing. Avoid one-eighty's if you can.

If you can't, and the river is narrow, then expect that you may have to angle towards your side, shut down, coast to the bank or shallow water, get out and swing the airplane, get back in, restart, and so on. (Keep your passengers in their seats, don't let them help, always hang on to at least one handling line during the turn-around, and don't hurry.)

With an amphibian, take care not to foul the nose gear mechanisms with mud/debris from the banks. If you do, wash them clean before takeoff.

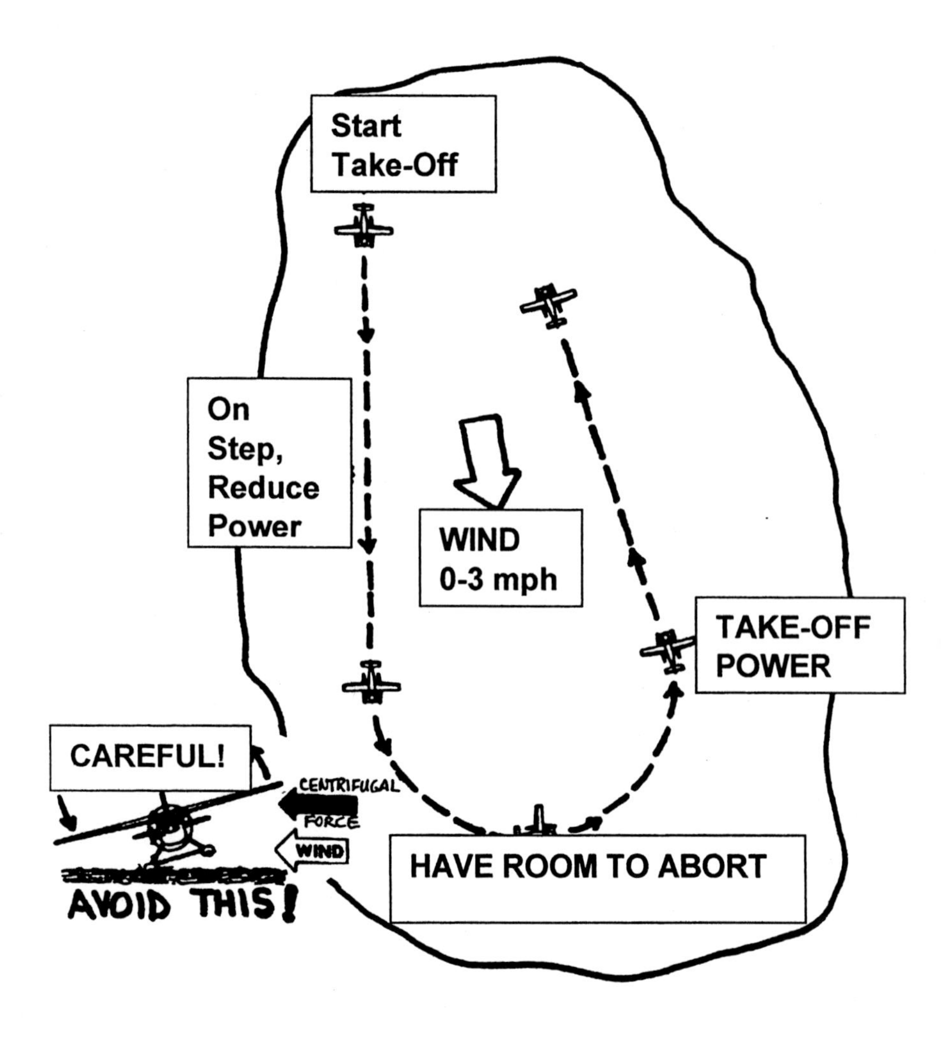

LIGHT WIND, SMALL AREA TAKEOFF
FIGURE 5-1

Small Water Areas

Before landing a float plane on a small lake or lagoon, evaluate the likelihood of your takeoff if the wind were calm or different direction. Consider your capability to climb above the level of surrounding terrain level while remaining over water. If not, does the surrounding terrain allow you to climb for safe departure?

For calm to "light air" takeoff (figure 5-1, preceding page), you may have to get on-the-step while going downwind and keep the plane just "over the hump" on-the-step. Then, with smoothness and care, use power to hold your on-the-step speed while turning into the light wind. You're staying on-the-step, yet going as slowly as you can to minimize turning radius, but yet you must keep the wings level.

When you have the right amount of turn remaining to reach takeoff heading, smoothly go to full power while continuing to keep the wings level. The "right amount" depends on your airplane, its load, and wind/water conditions – it could be zero degrees to go. Practice in large, light-wind areas before trying this in an actual small lake. Adding to your load will increase the centrifugal up-setting force during the turn, so practice with expected loads. Do not try this in winds above 3 mph!

You can know these factors only by practicing this type of low-wind turn into the wind with various loads in larger bodies of water.

Flight instructors correctly tell their float plane trainees not to turn into the wind unless the power is at idle and the seaplane is in displacement flotation. But, after experience with the fool-proof way, you may want to expand your operating envelope, carefully.

Allow no lean toward the outside of the turn. In fact, I avoid having to use full aileron towards the inside; then I'll know if I'm close to "running out of aileron" required to keep sure control of the wings.

Indeed, getting on the step downwind and then turning <u>into</u> the wind can be hazardous, even with no wind, if you're not smooth and careful. Make this sort of turn as wide as the lake allows.

If the aircraft gives the <u>least</u> hint of heeling outward, quickly and simultaneously reverse the turn with full <u>opposite</u> rudder, chop power to idle, and full up-elevator while holding full inside aileron, all to avoid a dunking. With some models, it will also help to retract wing flaps.

Also, with "<u>light air</u>" winds, a light-wing-loaded airplane, and <u>shallow-V</u> floats, consider the technique of taking off about fifteen to thirty degrees from wind-line and lifting the <u>upwind</u> wing while maintaining heading with rudder as you near takeoff speed. Don't try this with deep-V floats, a high wing loading, or with wind more than three mph. If you do, the drag increase on the downwind float will negate the benefit of getting the crosswind under your upwind wing.

With high wing loading, with deep-V floats, or if you don't have a lot of float experience, do not try the lift-the-upwind-wing technique. In any of these light wind situations, simply use the normal, safer, lift-the-downwind-wing technique for light wind, small lake situations. With both techniques, it's very important to keep the airplane straight with rudders during lift off, and to be smooth.

To repeat, if the winds are stronger than three mph, don't lift the upwind wing for a cross-wind takeoff. Rather, use standard cross-wind technique and lift off with the upwind wing down while holding your heading with "top" rudder.

High Elevation, Small Lakes

Try to avoid combining a small lake with high elevation. If you've got to try it, then practice high density-altitude operations, with more than adequate operating room, before you expose yourself to this challenging situation.

One way to get accustomed to high density-altitude operations is simply to takeoff at sea level with the throttle setting reduced from the normal sea level takeoff indicated manifold pressure by one inch for each one-thousand-feet that your prospective operating area is above sea level. If your plane doesn't have a manifold pressure gage (normal for a fixed pitch prop), get one installed or try reduced settings of the throttle. Doing this at maximum allowable gross weight and in calm or light winds will give you plenty of insight into your

airplane's high density-altitude capabilities. If your engine is turbo-charged or normalized, here's where you'll appreciate that feature.

Glassy Water

Glassy water is special and also should not be combined with a small operating area. If you are becalmed before takeoff, consider churning up the water with a few runs near the takeoff area just prior, but don't overheat your engine. Breaking suction by lifting the right float first (while holding a steady heading with rudders) helps. You just have to try glassy water with plenty of room and get to know your airplane/float combination.

Be super-careful when taxiing on the step in glassy water. You can inadvertently get airborne and fly back into the water nose-down, easily! If you do get airborne in this situation, use shoreline and flight instruments for reference and complete the takeoff and climb. Then, you can <u>methodically</u> use good glassy water technique to re-land if that is your intent. Avoid impromptu changes in such hazardous situations.

Another glassy water hazard on the step is not being able to judge your speed and therefore closing the beach too rapidly after landing. You'll do this only once -- guaranteed.

Regardless of the takeoff option you plan <u>prior</u> to landing, small areas require familiarity with your aircraft's slow flight characteristics as well as its

takeoff capabilities. Without confident expectations of your airplane's performance, reject small or questionable water areas.

Salt Water Float Plane Evaluation

The float plane pilot, when evaluating wilderness water, fresh or salt, should keep as a basic precept the goal of landing, taxiing, and taking off under the best possible conditions for minimizing shock on the float plane's structure. There is no merit in landing or taking off in rough water when smooth water is available and nearby in that same bay. Structural shock damage accumulates in the structure; that is why floatplane structural problems are usually fatigue failures. In that sense the plane's structure is a record of pilot indiscretions.

For landing in salt water, you will do well to monitor the swell pattern while enroute and evaluate its probable effect (surge) in the contemplated landing area. Sometimes that is difficult until you make a low pass. More often than not, you might want to pick a landing on swells at an angle in order to "stretch out the wave." This could give you a cross-wind and so, you'll want to be careful about getting "too parallel" to swells with steep peaks. The idea is to not allow a wind to get underneath the upwind wing.

In an area wide enough, you can start your airborne evaluation into the wind, at about 50 to 100 feet above the surface, at approach speed, and start a gentle turn away from the wind. Watching the

swell as you turn, you will likely know when you've "stretched" the wave enough for a less violent arrival. Look for the occasional higher than average swells. You'll want to use good rough water cross-wind technique (as required) to touch down beyond one of the higher swell crests. In a series of seven to ten swells there will be at least one or two that are higher than the others.

Sometimes, the long, low swells from distant storms are best detected from an altitude above about 1000 feet. Studying the swell pattern as it meets the shore will help. In light cross-wind, it will often be best to parallel these swells.

Strong Directional Wind Shears With Downdrafts

With strong offshore winds you sometimes find in bays, look for two wind patterns. First, you will have the lower surface winds coming from the low pass to create water waves and wind streaks that influence your planning. These low-level winds are shaped in speed and direction by the valleys that empty into the bays.

A second wind pattern, the higher, stronger, pressure gradient wind, often will show its presence as dark "williwaw" patterns on the surface. The stronger wind blasts down through the surface wind pattern and kicks up spray. This second, descending wind pattern is created by the stronger pressure gradient winds coming over the ridges and bluffs that confine a bay. Many times their direction is at a large angle with the surface pattern. That

directional wind shear makes things very sporty.

Large directional wind shears can rapidly change your airspeed in the air and turn or even flip you over on the water. (See figure 5-2.)

WILLIWAW WINDS
FIGURE 5-2

In this situation it's best to touch down, and to taxi, in areas least affected by williwaws. Your landing should be on the skegs with power on, as

opposed to being fast and flat with less dynamic directional stability. In fact, some floatplanes have very little positive directional stability during fast, flat touchdowns. In the flat attitude and with a sudden directional upsetting force, the plane can start disintegrating before you realize what's happening.

Study the williwaw pattern and locations. Then plan your departure, surface, and landing evolutions, so as to use the surface winds to advantage, while minimizing your exposure to the williwaws.

Rocks, etc., and the Seaplane Evaluation

Seaweed likes to anchor on rocks. Wherever you see fixed seaweed, likely they cover rocks. However, storms can rip seaweed from rocks, depriving you of its flagging feature. You can't beat a careful pre-landing inspection.

Always look for floating and semi-submerged logs in salt water areas, especially during spring tides (extreme highs and lows). The land above the beaches can be devoid of trees but adjacent waters can contain logs from over a thousand miles away. For example, in Pacific waters off the treeless lower Alaska Peninsula and eastern Aleutian Islands, there are always floating and semi-submerged logs from distant shores.

Know and observe the tidal status of any saltwater beach you visit. You need this

information to avoid getting trapped high and dry on an ebb tide. Try to arrive on the flood when you can. It makes things simple. Further, expect having to deal with tidal current, sometimes very strong, along any saltwater beach. A fresh water equivalent is the current near the outlet of a lake.

Evaluating For Land Plane Operations

With a land plane on typically soft off-airport surfaces, you can expect the takeoff to require more distance than the landing. The extra drag will always be there in soft snow/soil/sand/gravel to slow your acceleration on takeoff and increase your deceleration on landing. If you are inexperienced at off-airport operations, it is recommended that you start with three times the handbook numbers for takeoff length. Then, with experience, revise that as necessary for different surfaces, winds, and your particular airplane. With a firm, packed snow or hard-pack mixture of gravel, rock, and fine sand or soil you sometimes will get performance close to paved runway numbers.

On bare ice with skis, beware: the seemingly never ending run-out coupled with loss of directional control in light winds (and when below rudder effectiveness air speed) will produce surprises and hazards you'd best avoid. In this situation, be sure and pick a level spot for your tie-down. Then, plan how you'll get there after landing, giving due regard to wind effects on your ability to steer the airplane. Taxiing into the wind as you approach your parking area is best.

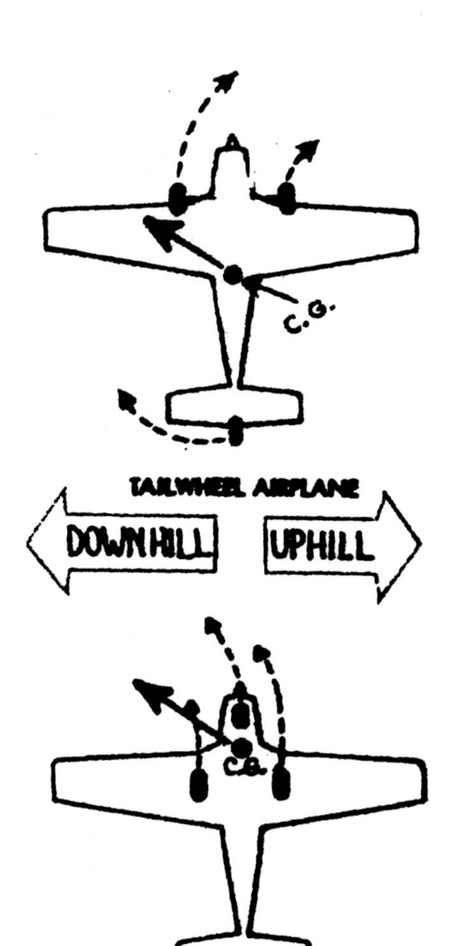

ENTERS OF GRAVITY ON SIDEWISE SLOPES
FIGURE 5-3

For wheel planes on side-wise tilted surfaces like beaches and cleared areas along hillsides, the

direction of even a light wind is very important. For the <u>nose</u> dragger, you'll need a wind parallel to the beach or strip. That's just the best available compromise. This is because, with an onshore (uphill) wind, the nose dragger will weather vane as well as be pulled downhill by its forward CG. (Figure 5-3, preceding page, exaggerates the C.G. locations but is correct on relative positions.) But the nose dragger can tolerate light onshore wind with shallow beaches.

This same onshore (uphill) wind will be just right for the <u>tail</u> dragger, which wants to turn <u>uphill</u>. In fact, for the tail dragger, the stronger the onshore wind, the better. For both types, the onshore (uphill) wind fits in with wing-down cross-wind technique to have the wings more nearly parallel to the surface. You should get your cross-wind wing-down correction established before getting near the surface, to avoid dragging the uphill wing tip. On takeoff, delay leveling your wings too soon, for the same reason. Due to your habit patterns with civilized runways, you'll have to remind yourself on this as you start the takeoff roll.

For both nose and tail draggers, an offshore (downhill) cross wind is "bad news." It's difficult to counteract drift without digging a wing tip. One has to make wings-level skidding turns to correct and that's unacceptable just above stall. Furthermore, in the typical narrow landing area, it's a challenge to swing out of the crab without picking up downhill drift prior to touchdown. The downhill cross-wind, no matter how light, is a good

reason to reject a site, regardless of how skilled you are with stick and rudder.

After considering the wind, from about 500 feet you'll want to ask questions like: Is the area long enough for an easy takeoff and climb-out with the expected load? What and where are the nearest obstacles to normal approach and climb-out? Are there any doglegs or curves just beyond the expected touchdown or just before the departure liftoff point? These are the points where you can not afford to turn because of greater centrifugal upsetting force at those higher speeds.

Measuring Your Prospective Operating Area

Timing your flight along the length of an area is certainly useful when you're gaining experience in off-airport operations. At 60 knots TAS, you're doing 100 feet per second, no-wind. At 75 knots it's 125 feet per second, and at 90 knots it is 150 feet per second, no wind. Timing into the wind, if you can compensate for the wind speed by adding to your airspeed, the result will be more useful in estimating the area's length.

Averaging times into the wind <u>and</u> downwind can mislead. It tends to <u>over-estimate</u> the length. That's because you spend more time into the wind than downwind, and with more time for the wind to work as a head-wind, the average is weighted towards the longer timing. (It's similar to the problem of mistakenly thinking that going on a

cross-country from A to B and back to A doesn't take any more time or fuel <u>with</u> a wind blowing than it does no-wind. It does take more time when there's a wind blowing. That's because the head-wind reduces ground speed and gets <u>more time</u> to work its adverse ways on you and the tailwind gets less time in which to help you because of your higher ground speed.

Measuring an area's length is good for reassurance when you are just starting wilderness flying and need to "calibrate your eyeball." However, once you gain experience, the use of timing to decide on a marginally short area is not a good practice. If you question the length of an area, that's good grounds for you to reject it. It's very difficult to predict how soft the ground is or how sticky springtime snow is, and therefore how "draggy," an area will be. The practical difficulty of surface drag is how it lengthens the takeoff run. (The nose-up problem associated with soft surfaces is covered in another chapter.)

Wildlife

From 500 feet AGL, look for wildlife. Having a band of caribou or other deer stroll onto the landing area just as you touch down is a thriller. If a bear is there and doesn't leave when you arrive, don't try to spook him away. That's usually against the law and it likely won't work. Just go to your alternate.

Often, simply making a low pass over the landing area with landing lights on will clear the nearby area of wildlife. One never knows. Wild animals, just like human hikers, like to use flat, open areas to save time. So do cross-country skiers, dog mushers, and other off-road travelers. Expect to see them in wilderness areas. When near fishy creeks, expect to see plenty of bears fishing, or hiding in nearby brush as a result of your approach.

Beaches

Ocean beaches are the chameleons of off-airport sites. They will change for the worse very quickly, even in one twelve to thirteen hour tidal cycle. For example, a friend once departed a good ocean beach in the morning. He returned later that day on the ebb. He didn't evaluate the beach prior to landing. Upon landing, a newly exposed part of a large crab pot removed his landing gear. His wing tip dug in and the wheel plane disintegrated around him. He survived with minor injury to learn a cheap lesson: No matter how recently you left it or how experienced you are with that beach, always look it over before landing.

When evaluating saltwater beaches, concentrate on the area you ordinarily will be using. That's the wet, hard-packed area just below the lowest high tide line. Generally, the dry part above the highest high tide line is fluffy, bumpy, and full of hidden obstacles. Down in the landing area, look for rocks, semi-buried logs, clumps of seaweed, fishing net lines, and hard-to-see undulations caused by storm

waves. These surface undulations can put you into the air in a stalled, ballistic trajectory during landing roll out if you encounter them when fast but below flying speed.

Look along the upper dunes for drainage notches. Then look below each notch for evidence of streams that run under the upper beach and emerge in shiny soft spots. Those soft spots will grab your wheels during roll out. In fact, any area that's shiny should be avoided.

Avoid combining the short field situation with beach operations. It can be very sporty, and you will wish you hadn't - especially when it comes time for the takeoff.

As the common ocean beach gets closer to confining points and headlands, the curve of the beach will tighten, the beach usually gets rockier, and the rocks get bigger. Look for this, in order to avoid major surprises on your landing roll or during takeoff later.

With spring low tides (when the low tides are the lowest), you might be tempted to land on the flatter area just above the water at low tide. Don't do that. The lowest area will have a lot of concealed soft spots, water-logged timber, and other things too dense to be washed up on the high areas.

Be most skeptical about saltwater lagoon beaches. They're usually narrow and sedimentary, with more soft spots than the usual ocean beach

which is frequently "scrubbed" clean by storms and big tides.

Beaches made up of volcanic pebbles will usually be steep, soft, and fluffy. That makes takeoff difficult without a strong, onshore wind – say, at least fifteen knots or more. With a heavy tail dragger on big tires, twenty knots onshore is a better minimum.

Look for bears near salmon streams and caribou standing in shallow water to escape insects. Caribou will usually stay clear; a lot of bears won't, especially the local "alfa" males. These large males are trying to defend their “personally owned” salmon stream.

For several years I was tormented by a male, ten-to-eleven-foot Brown bear at a Pavlov Bay salmon stream on the Alaska Peninsula. It always happened during the late spring chore of erecting State Fish and Game regulatory signs on salmon stream beaches. This huge curmudgeon would lay in a depression of the deep grass about 200 yards behind the dunes while I did my pre-landing evaluation. Every time, I hoped he would stay put, like most big males would do. He would, until I was near touchdown on my landing attempt. Then he'd come galloping towards the plane. I’d have to make a go-around and depart for the next beach on the list. I always hoped the local bear guides would nail that trophy-class hide during bear season. They didn't; he was just too smart.

Finally, I started waiting until bear mating season in June to put up the regulatory fishing signs on that particular stream. Then, while he was away doing his job, I'd do mine.

Evaluating For Ski Operations

Examining off-airport snowfields from the air for skis is like evaluating a milky river for float operations. In both cases, you don't know what your airplane is going to hit or sink into on landing. Snow can cover logs, rocks, stumps, and holes to make a field look smooth and safe. That's why you're almost always better off using a frozen lake covered by snow. Lakes don't ordinarily have debris on the ice under the snow, except maybe near a beaver lodge.

But they might have a lot of snow-covered obstacles if people live or play nearby. Lakes are flat, which is good, but they can have hard, hazardous snowdrifts.

If you can find wind swept areas to see the ice, and you'll definitely want to when air temperatures are up, you usually can get an idea of its thickness by its color. White is ordinarily thick ice, and gray or dark is dangerously thin.

Overflow, The Scourge of Lakes

Lakes are convenient but, in addition to drifts, they might have an affliction that can trap you: overflow. Overflow is usually an ice-water matrix

(slush) on top of the ice but insulated from the cold air by a blanket of snow. Overflow can be any depth from a few inches to several feet. It can be standing or flowing. When uncovered it's surface can freeze into a thin, rough surface.

Overflow seems to come from three sources. First, there can be a heavy snowfall early after freeze-up before the ice gets thick. The weight of the snow cracks the ice and the combined weight of ice and snow forces geysers of water up through the cracks. Second, lake ice with a thin blanket of snow or areas of no snow can crack during quick, deep cold snaps, releasing water. Third, a lake can get runoff from year-round flowing springs, or during thaws, from creeks flowing onto the lake ice.

Overflow can be anywhere: in the center, along the shores, or all over the lake. As one might expect, overflow is most likely near creek or spring outlets onto the lake from shores and islands. It can also be found near a lake's ice-blocked outlet. Further, you can't count on any particular lake being the same every winter.

The <u>sequence</u> of weather events seems to make the difference with overflow. The sequence that minimizes overflow appears to be starting winter with a long, gradually deepening cold spell with steady, accumulating snowfalls with no temporary thaws.

Your best way to detect overflow is on snow shoes. That is, if you can, have someone at your

destination check out the lake you're considering and put out a single line of markers, along the wind line, where the ice surface is dry for an appropriate distance either side. Even a "little" water is disqualifying. The overflow could be just starting or be deeper than the reporter thinks.

By now, if you're thinking you will never volunteer to be the first one (or the last one) to land on a particular lake that season, you're getting the right idea.

Evaluating for Overflow

Even with an advance report from someone at the lake, do a thorough evaluation from the air. Evaluating from the air for overflow is a lengthy process.

First, and this will take some practice, do a preliminary run with just the heels of your skis, but without settling fully onto the skis' axles. That requires a hanging-on-the-prop attitude, so be a little faster in rough air. Then, climb out and quickly look to see if your ski marks turn dark from unfrozen water. If they don't, it is so far so good.

Next, make a run in the same area, putting more weight on the skis while keeping your airspeed up and takeoff flaps set. Then take a look at the fresh marks. You should avoid getting too slow or trying fancy high-speed patterns. Patterns will slow you down as you turn out of the wind. If you get slow and run into overflow, you're stuck. Making several

touch-and-go ski runs parallel to and in your intended landing area usually will suffice. Slowly build up the amount of weight you put on the skis during your passes. (Plan on extra fuel for all this.)

You should look at the surface immediately after each pass. Otherwise, the dark color of any water will disappear as it freezes and you won't see it.

Taking your time, plus being lucky -- these are the keys to avoiding overflow. Some things are never guaranteed, and not getting into overflow is one of them. Because of this, the wilderness ski-plane pilot is well-advised to carry an ax, a couple of saws, a shovel and some extra rope, in addition to the normal camping and survival gear. (Payload will suffer for a good cause.) Then, if you get stuck in overflow, you'll have a better chance of extracting the airplane, making a shallow ramp built of small poles placed across your intended route of extraction.

Thin Ice

Overflow is difficult enough, but settling through thin ice after landing is to be avoided at all costs. If lake ice is any color but white as snow, don't even consider it for a landing. As long as the lake ice is not near open water and not starting to develop cracks, the color test will work for smaller aircraft. We're talking about bare ice. If you get snowfall in the spring after the creeks are running, it's best to stay off the ice without on-the-ground daily monitoring of ice thickness.

It's good planning to stay off of frozen creeks and rivers all winter long, unless you can have up-to-the-minute ground-based reports on specific landing sites. The ice color test is not reliable for creeks and rivers, or wherever there is running water such as near lake outlets . Ask anyone who has hunted along creeks or rivers in below-zero weather. The ice can be white, blue, gray, or dark without a consistent thickness tied to particular colors and shades.

When you are looking over an area for ski operations, try looking at it from different directions. The different lighting angles will help you see rough areas or breaks in the surface. The more you look, the more you're likely to see. Keep circling and evaluating as long as you keep seeing new things. Sunglasses with polarized lenses can be helpful on that. When you have two or more broken layers of cloud above you, the lighting will sometimes change during your evaluation due to relative movement and coincidence of holes in the cloud layers.

Ice Fog

Another pertinent weather phenomenon occurs when the air temperature is below about minus fifteen degrees Fahrenheit. Because water is a product of internal combustion, you could unknowingly be laying down a "frozen smoke" screen from your exhaust stacks. So, check your “six o'clock” as you approach the landing site and

do your best to evaluate from the downwind side if that is required. In that case, be ready to say "no" to a site if you can't give it a good look-see due to the ice fog.

White-out

Avoid flying low over a large white expanse; it's a perfect setup for "white-out." If you must land in such a place, check your altimeter for reference as you leave nearby areas with visible vegetation or rock. Having brought along pieces of tree boughs, slow down and toss out a piece every three or four seconds along the wind line of your intended landing area. It's messy, but very worthwhile because it turns an unsafe situation into one that's manageable with methodical procedures. Dark, weighted, and biodegradable plastic bags are good for this, too.

In any place, including areas not so wide, be alert for white-out if things on the surface look fuzzy or if it is snowing.

Surface Wind and the Ski Plane

Because of the ski plane's limited directional steerability, you need to pay close attention to wind direction. If you land downwind in a wind of ten knots, count on losing directional control at the usual speed plus ten knots. If you must taxi downwind, expect to end up having to travel at something faster than the surface wind.

Always have a alternative plan if you lose control. This plan will sometimes include shutting the engine down or it might include a blast of power coupled with full rudder and a one-eighty. Planning to land or even taxi downwind without such an "escape plan" is inviting disaster. Generally, it pays to land into the wind, even on shallow downhill landings, provided there is adequate level area beyond for the run-out.

Some techniques of maintaining a semblance of control with skis are given in the next chapter on landings. Usually you should reject any ski landing area which has side-wise, tilted gradient but, if it looks level enough for your particular airplane, plan carefully your landing run-out, taxi routes, and takeoff/departure procedures before you land.

CHAPTER SIX

LOOK-SEE APPROACHES WITH FINAL APPROACH AND LANDING

This chapter completes the off-airport site evaluation, followed by the landing itself. Included is an appendix narrating the typical off-airport landing from the hands-on-controls view.

In flying, as in many pursuits, there is more than one way to "skin a cat." The proof of the pudding for you can be the realization that what you're doing in the pilot's seat is working and you're not banging up airplanes or hurting people.

The Look-see Approaches

To complete your evaluation of a little-used airport or an off-airport site you should make approaches for a closer look, each followed by a go-around. I call these “look-see” approaches.

Making look-see approaches confirms and discovers. You confirm what you thought you saw at higher altitude and airspeed. And you often discover things you had not see.

Your evaluation is proven successful only upon the safe completion of your landing <u>and</u> departure. I admit to having seen some important obstacles only after shutting down and debarking. That's where being lucky helps.

Outcome Risk Analysis

"Outcome risk analysis" is a good basis for choosing a course of action from more than one option. The choice is made primarily on the basis of the probable, acceptable, "least risk" outcome for each available option if that option does <u>not</u> work as planned.

The choice should <u>not</u> be made on the basis of what will most enhance the pilot's status, due to the risks involved. The bush pilot who expects to live long will always minimize the risks, knowing full well there are plenty of risks inherent in what he or she is doing. In other words, it's the use of plain old common sense. Outcome risk analysis can be illustrated by the following two examples.

Risk Analysis Examples

While on a survey flight along the Bering Sea coast of the Alaska Peninsula, the author saw a familiar Cessna 185 on wheels approaching a tiny strip. I had used the strip in a Super Cub and was familiar with its hazards: lumpy tundra on three sides and a high cliff dropping to the ocean beach at one end. Wind was calm. Unbelievably, the 185 was landing slightly up slope, toward the ocean. I confirmed later that the up slope drove the pilot's decision. The truth was that the strip was just too short for that particular model in a calm wind.

Undaunted, the pilot even landed a little long,

stayed on the surface, and terminated the roll-out with a one-eighty ground loop in a swirl of dust about thirty feet from the cliff. The plane tipped, almost dragging a wing tip, during the ground loop. Obviously, the "loose cannon" at the controls either didn't know or care about outcome risk analysis prior to landing.

As an FAA Aviation Safety Counselor, I was obliged to have a "heart-to-heart" with him later when our paths crossed. He agreed it was a big lapse in judgment with small penalty. We talked of outcome risk analysis and parted still friends.

For a positive example of outcome risk analysis consider the following hypothetical, yet frequent kind of situation.

You are taking friends plus their gear into a valley with a typical broad, braided river in the interior of the Alaska Range. This is your first of several loads. You're in a Super Cub with all the usual "Alaskan mods" including large wheels. Right now you're evaluating a smooth gravel bar stretching between two dry channels as your staging area. The wind is calm but the length is adequate for both takeoff and landing in either direction.

The surface slope is very shallow. That's why the river is braided. Slow glacial, silt-laden rivers fill up their own channels. Then, they overflow Into new routes.) The up-river end of your chosen gravel bar terminates with a steep, four feet cut bank. The down-river end leads gently into the smooth curve

of a wide, shallow channel. Both channels are dry.

You do a quick outcome risk analysis. Although you usually feel more comfort with an up-river landing, you decide that it's best to land and takeoff down-river. That's a good decision because if you land a tad long or you have to abort late on subsequent takeoff, you're much more likely to have a better outcome in a down-river direction.

In this case, the sharpness of terrain changes at the ends of the operating area makes the difference. Going off the up-river end, you'd have a sharp drop over the bank onto your prop. Down-river, you'd have only to turn smoothly as you come to a stop in the dry channel.

Outcome risk analysis, good judgment, or common sense -- call it what you wish -- just exercise it every time, as a conscious part of your off-airport evaluation routine and you'll be a much safer bush pilot.

Making Look-see Approaches

The purpose of look-see approaches, as stated in the opening page of this chapter, is two-fold: confirmation of what you've seen and discovery of new things. Often you'll need only one look-see approach. Do at least one, the situation permitting. Sometimes you'll need more. Don't rush.

Evaluating One-way Landing Areas

For a one-way landing area, you need to be creative in choosing a look-see flight pattern. Obviously, the look-see pattern cannot be the pattern you'll use for final landing, because of obstacles or terrain. For most of these areas, flying a reverse direction or crossing the landing area at ninety degrees, at safe altitude will be adequate.

If you are considering going into a one-way strip that does not permit a look-see (usually due to weather), don't do it. The author has been asked to do that sort of thing, but declined. They were leaps of faith much too big. A report from an on-site, knowledgeable pilot would justify giving it a reasonable try.

Emergencies

Urgent requests that ask you to violate your usual precautions or personal wind/weather limits will test your resolve. Taking off in order to "check it out" can lead you to violate your own minimums. If you do takeoff to "check it out," go alone. The author recommends that you adopt the following operating policy for all emergencies, no matter how urgent:

"Emergencies can justify a change in flying hours and priorities. However, response to emergencies must never violate the minimum requirements for safe flight. One emergency should not be allowed to create the potential for another."

The First Look-See Approach

During the first look-see approach, pick a landing glide path aim point. That should be a clearly visible object or discoloration, on or at the edge of the landing surface that you expect to use for the final approach to landing. Also, pick a second reference point, alongside the landing area, that is halfway to the far end of the landing surface.

If you're uncertain about the softness of the surface, the acceptability of departure obstacles, or the severity of wind gusts, fly the look-see approach at higher airspeed than normal.

If your concern is for some existing departure obstacles, don't make a touchdown on the first look-see pass. Take the time to settle that issue before touching down on the prospective surface.

A slightly higher airspeed might be necessary for a crisp recovery from touching down on a too-soft landing surface. As you fly down the glide slope look at the far end of the area to verify the suitability of the departure area. Study the length-wise slope and side-wise tilt of the landing area. Be alert to the strength and direction of the wind and possible wind shear.

It's not unusual to discover a sloping "runway" on your first look-see. It could even be so steep that you reconsider your original plan. With light winds and no obstacles, you might want to land uphill and takeoff downhill.

Don't get committed to a particular decision until you have seen everything, and apply some outcome risk analysis (common sense).

The Tail Wheel Index

Use the radius of your tail wheel as an index of vulnerability to surface damage. The main wheels will roll over things and keep you from sinking in but the tail wheel is much more vulnerable to rocks and debris as well as to sinking into soft surfaces.

Rocky areas made up of loose, non-imbedded rocks bigger than three to four inches should be rejected as too rough for the typical eight-inch tail wheel. Even though you use the technique of holding the tail wheel off the surface as long as possible on landing roll-out, the tail wheel should be the index for safety on a rough surface.

Other Surface Hazards

Look for undulations of the surface. These can be difficult to see in flat light with no shadows. Sometimes, on a flat light day, you can see hazards better if you make at least one downwind look-see pass. For example, when the beach has rocks the same color as the sand, or when there are

hazardous undulations of a beach. Again, look for obstacles along the full length of the landing area. Check the vegetation along the sides for "sweepers" sticking out far enough to grab your wing. Dark, leafless alder branches don't show up well against dark-colored dirt or sand. You must assume that the branches stick out farther towards the open beach or strip than they do in other, more restricted directions. Also check for holes, depressions, and shiny soft spots you may have missed in your initial evaluation from four to five hundred feet AGL.

Check the surface for tracks of wild animals, humans, vehicles and airplanes. With six to eight psi in typical bush-flying mains, you should be able to land safely on shallow animal tracks.

But don't be fooled. I once found airplane tracks on a snowy village strip and saw that the tracks led to the standard-size main wheels of a Cessna 185 parked by the snow-covered runway. I then made the false assumption that the 185's wheels had made the tracks and thus the runway snow was hard-packed. Therefore we could land safely on our small Cessna's wheels. Wrong! With my flight student watching closely, I landed in <u>deep</u> snow and nosed-up to make a slight (but $170) bend in the propeller.

You guessed it. The 185 owner had just shifted from skis to wheels. After that, he had tied the 185 down to wait several weeks until the snow melted and "muddy breakup" was over. (I never saw that

student again, and I didn't blame him.) This mistake would have been prevented if I had called the Flight Service Station and checked on the village runway's snow condition prior to flight.

Sometime in the future, you could arrive over an off-airport area you've previously used and find it covered with wild grass. If you can't see down into the grass well enough to ensure that it doesn't hide obstacles, don't land there. Sadly, during hunting season, I have found such operating areas blocked with almost hidden rocks and logs to prevent others from using them. Look out for this one!

The Touchdown Test

If you are reasonably confident that the surface is acceptable for a touchdown test, plan on a brief touch-and-go. If you feel rushed and not yet sure that touchdown is a good idea, go-around. Make another pass to clear your uncertainty. As you near the flare for touchdown, look at the far end of the landing area for the purpose of determining whether or not there is enough distance available for your later takeoff, your planned full stop, or even a brief touch-and-go on this pass. If it looks too short, it is. Reject the area in this case.

Keep your speed up, with some power on, and plan a wheels-type of touch-and-go at only slightly higher than normal touch down speed. However, touch down in a location you plan to use for your full-stop touch down. Be ready to go around, no matter how close to the surface you are, if you get

an uneasy feeling about anything. Trust your instincts when you get a negative feeling about a place.

The Big Surprise - What To Do

Upon touch down, if you feel a sudden, big, nose-down tug, quickly go to full nose-up elevator, adding takeoff power to get airborne. Once you get off the surface establish the appropriate climb attitude and power. The surface is too soft and must be rejected.

If the airplane remains stuck on the surface, keep the elevator at full nose-up and maintain full power. Then, gradually reduce power. If, at any time, the nose starts to drop, quickly increase the power and keep the stick or yoke at full nose-up position. Look for higher or drier ground and try to steer the plane with rudder (no braking!) to reach that area. As the plane comes to a stop, you should smoothly reduce the power to avoid getting a prop strike on the surface.

The above technique might <u>not</u> work if your airplane is lightly loaded or has a forward center-of-gravity. If putting on full power at the first big tug doesn't get you airborne and, even with full nose-up elevator, the airplane continues to go nose down, then your best move is to chop the power quickly and secure the engine.

The moral of this story is that when you are light or nose-heavy, don't commit to a soft or soggy

surface. Even if you are heavy, avoid it.

Fortunately, the above technique also will work on skis, in the situation where the snow/ice was thinner than you thought. Consequently, as you slow and loose lift, you break through to dirt/gravel and get that big tug from surface drag.

It also might help on wheel-skis when you land with the skis in the wrong position. It depends on the situation. If, however, you settle into deep slush (overflow) on a snow-covered lake, you will stop quickly and are better off to reduce power as that happens. In any of these ski situations, the cables and fittings to the heels of the skis will have been stressed heavily, and so you'll want to check them for damage before flying again.

To reiterate, in any high surface drag situation, if increasing power and holding full nose-up elevator will not keep the tail down, reduce power and shut down. You will come to a quick stop regardless and therefore should avoid a power-on nose-up in order to possibly save the prop, engine, and accessory drives.

Completing the Look-See

To continue with your touch-and-go on the look-see pass: stay on the surface but keep your speed up at just below lift-off airspeed. Somewhere between one-third to one-half of the way down the "runway," you should verify that you will be able to

takeoff and safely climb-out on departure. <u>Before</u> you reach the point where safe climb-out becomes questionable, leave the surface, get on appropriate climb speed, and start your climb.

Note the point where you lifted off. If that "questionable point" is short of the halfway point, you likely do not have adequate length for safe operations. In other words: you'll probably need half of a minimum safe available takeoff run just to get up to liftoff speed. Climb obstacles and wind strength and direction will be major determinants. And, of course, the softer the surface, the longer is the required takeoff distance.

When you complete this evolution, you should know whether or not the landing area is suitable for subsequent takeoff. Uncertainty itself is grounds for rejecting the area -- this not the time and place for hopeful guesses or wishes for more wind.

Never plan to use wheel brakes to compensate for a short or questionably short landing area. If you do, sooner or later you will come to grief with a nose-up or worse. Even with impeccable stick-back and alternating sides braking, you simply cannot predict how quickly the wheel brake pads will release each time you come off the brake -- it varies in large part with the resistance being given by the specific surface the wheel is on at a given moment.

For example, on a dry concrete runway, the brake pads will become ineffective as you release brake pressure and the wheel quickly will resume its

unrestricted rolling. At the other extreme, on a soft, loose surface there is much less surface friction available to cause the wheel to spin-up or to keep the wheel turning. Therefore, the wheel can "build chocks" with the least provocation from the brakes and cause a nose-up or, worse yet, a flip-over.

Consider the tug you get on big wheels from the high friction spin-up when you make a "wheely" on gripping runway surfaces. When making a wing-down cross-wind landing on gripping runways with big tires you'll get a big yaw when the low wheel touches. Take care not to over-correct because the second big tug from the other wheel is yet to come. Compare those big spin-up moments to the faint, minimal spin-up tug, if any, when you touch down on off-airport or gravel surfaces.

Because of the above, you want to minimize built-in wheel drag in order to minimize the possibility of "chock building" by the tires. For that reason, whenever I'm going into a spot that I know or suspect is soft, I'll take the time to jack up each main wheel briefly to see that there is no dragging of the wheel, and yet the brakes are effective when applied. Little things do add up.

Hazardous Rain and Light Conditions

During off-airport evaluation, if it starts raining, expect that it will interfere with your ability to see through the windscreen and safely land. In that case, reject the landing area until the rain shower goes through. If the rain persists, it's a good idea

to forget about landing when you find the rain drops rolling up and stopping on the windscreen as you slow to approach speed. Each drop becomes a bright spot. This can be very dangerous, depending on the lighting and the darkness or contrast of the surface. In many cases your best course of action is to head for a place where there is no rain.

The Look-See for Water Operations

In the rainy situation with a floatplane, the rain may or may not be a factor in your pre-landing survey, depending on its severity and that of the waves, but it could limit your ability to see surface and sub-surface obstacles looking ahead.

After the initial evaluation, the seaplane pilot will usually see more by looking vertically, with approach speed flybys at about 200 to 500 feet above the surface. Any lower altitude will have possible obstacles in the water going by your field of vision too rapidly to assess. Also, any lower altitude could be unsafe with glassy water conditions.

With some wind, a look-see approach down to about twenty feet altitude can prove useful to evaluate cross-wind and down-drafts. However, the water surface usually will be showing what's going on with the wind.

When the wind and waves are up, it is difficult to see rocks, including those projecting a little above the surface. Even at 500 feet AGL, it's tough. In a

river situation, with high wind, you can miss surface "humps" that usually pinpoint shallow, subsurface rocks and upwelling (flat, wide, smooth interruptions of the surface) that signal deep areas just upstream. Therefore, with wind-blown surfaces, you must pinpoint water obstacles and nearby reference points from at least 200 feet AGL. But, making a look-see pass down to about twenty feet will help in judging the height and width of banks in a narrow river.

Also, at this lower altitude you can get a better estimate of the lengths of straight river sections being considered for subsequent takeoff.

The saltwater seaplane pilot will be interested in examining the shoreline for evidence of tidal current. Any tidal drift can make it difficult to stabilize the seaplane on the beach for unloading and loading. Drifting foam, flotsam, and seaweed are good indicators. (Flowing lake water near a lake outlet creates a similar situation.)

If you observe waves and/or swells breaking on an ocean beach, do not attempt to use the beach or to beach the plane. If you do, the airplane could veer broadside to the waves (broach) after you arrive at the beach. Then, the waves will push the airplane up on the beach, pounding its bottom unmercifully. This, despite your best efforts to keep it pointed into the wind and waves. A large angle between wind and waves makes things worse.

Furthermore, any previously undetected tidal

current along the beach will force you to swing the nose of the plane somewhat into the current to compensate. When you do that, the wind and waves could seize the opportunity to broach the airplane. Having to deal with debarking/embarking passengers or unloading/loading cargo complicates all of this. Even the larger multi-engine seaplanes can have real problems in these conditions, and you will be less able to control their heavier weight.

As mentioned previously, be alert to detect swells and avoid exposing the airplane by using sheltered water and/or choosing takeoff and landing headings that will "stretch" the swells, wind permitting.

Conflicts With Boats

Finally, the water pilot must anticipate any capability of boats to interfere with plans. On the water, your seaplane is considered just another vessel from the standpoint of "rules of the road." You not only must observe rights-of-way, but you also have to focus on a boat's capability to interfere, with or without rights-of-way, and not assume particular intentions of any craft. A boat coxswain deals in two dimensions and may not even see you approach the surface to land.

Making critical assumptions of someone else's intentions gives up control of your own destiny. Further, avoid situations where you suddenly become the "burdened vessel" (required to give way to the "privileged vessel" on your right) upon completion of landing or during your takeoff. Try

to avoid putting a boat in the burdened position during those procedures. When privileged, technically, while on the surface you must hold your course and speed. You can't while landing or taking off. Thus, being "privileged" is no privilege. Moreover, you lose control of your immediate future by having to <u>depend</u> on the other vessel (boat or airplane on the water) to do right and give way to your vessel. Hey, lookout for boats as well as other seaplanes!

Often, one more circle in the air before landing or one more circle on the water prior to takeoff will eliminate the uncertainties regarding other vessels in or approaching your area.

The Off-Airport Pattern and Final Landing

A traffic pattern altitude of 400 to 500 feet AGL is appropriate for the bush pilot who navigates routinely at those altitudes. The idea is to set up for a shallow, power-on, precision approach. If you are accustomed to 800 to 1000 feet traffic patterns, the lower altitude will take some getting used to. When you're on the downwind leg, you'll have the urge to get much closer abeam. That's because you are accustomed to seeing the landing point at a certain angle below the horizon, much lower than what you will see from the proper distance abeam at 400 to 500 feet AGL. (See figure 6-1.) In hills, 400 feet looks even lower. (See figure 6-2.)

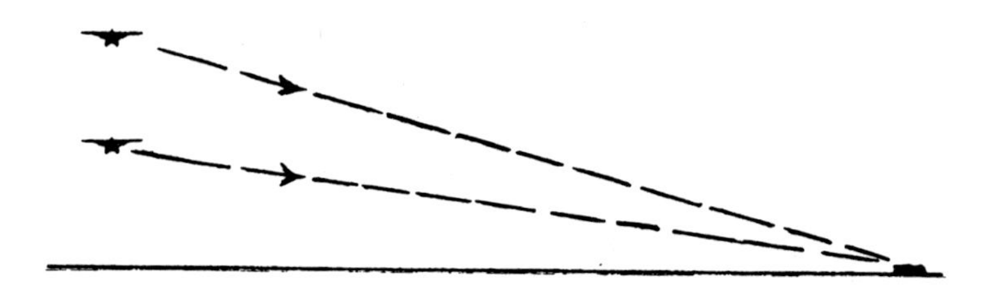

400' AGL VS. 800' AGL TRAFFIC PATTERN
FIGURE 6-1

400' AGL TRAFFIC PATTERN
FIGURE 6-2

One way to assure that you are at normal distance abeam is to cross the landing area at ninety degrees, fly straight for twenty seconds at about 1.4Vs0, then make an approximately twenty degrees banked turn to the downwind leg. (You'll have to adjust this for the wind at TPA.) In sum, it's: cross the runway, fly about twenty seconds on the crosswind, then twenty degrees bank to downwind.

Pick a downwind leg that best avoids obstacles both on downwind and on approach. For that reason, all else being equal, I use a downwind over water for beaches and the down slope side for a contour strip on slopes. On the downwind, after completing the pre-landing checklist, it's a good time to brief your passengers. The author explains to his passengers: although he plans to land on the next pass, the approach and landing could very well become a go-around if something new and significant to safety is seen.

The appendix to this chapter goes into detail on how the author makes a typical wheel plane landing in the bush environment. It will become clear to the reader that my emphasis is on picking an aim point, avoiding late line-up, intercepting and maintaining the glide slope with pitch adjustments, and maintaining approach airspeed by coordinating with appropriate throttle adjustments. I believe this makes it easier and safer to fly a constant approach speed down to the point where, depending on the situation, a partial power reduction is made in preparation for the flare to landing.

The alternative method of trying to slow down gradually with pitch adjustments while attempting to control altitude with throttle just doesn't work except in a calm, benign environment or in a fixed throttle situation. Corrective actions take too long to effect the desired changes. They are too indirect. Therefore, precision can be elusive in turbulence. (Of course, pitch must control airspeed in the two basic fixed-throttle modes: full throttle and engine

at idle. But here, we're trying to do power-on, precision approaches.) In sum, pitching to control airspeed and adjusting throttle to control altitude is not conducive to precision. In rough air, this method becomes a real joke and is hazardous. Observations as a flight instructor and pilot examiner have reinforced that opinion.

After touchdown, if the situation looks good, final commitment is made by reducing throttle to idle. The tail wheel is held off as far as possible during the landing roll. The nose dragger pilot will want to hold the nose wheel off the surface as far as possible. The idea is to protect your smallest wheel as much as you can.

During the landing roll, if the area is tilted sidewise, be ready to counter center-of-gravity effects. (These were explained earlier.) In this, you may have to use blasts of power combined with full rudder. Avoid brakes, but use quick, brief jabs to one brake at a time if really necessary. Sometimes you'll need to use elevator when appropriate to relieve pressure on, or drag from, the small wheel.

APPENDIX A TO CHAPTER SIX

DETAILED DESCRIPTION OF A TYPICAL OFF-AIRPORT LANDING APPROACH

Okay, we're on a 400 feet AGL downwind after having done the evaluation and look-see passes. The wind looks good -- about ten knots and thirty degrees left of the old mining strip where we are landing. I did my before-landing checklist (posted on the instrument panel of this ancient Piper Super Cub) before I made the first look-see approach down to the surface. But one more check is always appropriate prior to the final approach for landing. I remind my hunting partner in the rear seat to cinch up his seat belt and shoulder harness, in that order.

We are still below Vfe and, as I pass abeam my intended point of touchdown, I check to ensure we have twenty degrees of wing flaps for landing. There are no obstacles to my landing approach on a one-and-a-half degrees glide slope. So, I won't use full flaps on this approach. If I don't need full flaps, I prefer to fly the aircraft with less drag. In addition, as with most high wing airplanes and many low wings, this airplane's rudders seem more effective with less flaps.

The aim point, which is now approaching thirty degrees aft of my left wing, is just beyond a small gully cutting across the approach end of the landing area. We will touch down about two plane lengths beyond the gully. The landing area is a tilted strip along a creek. Although there is a tilt to the left, the

strip has no big up or down gradient, end-to-end, and is plenty long for our future takeoff at maximum allowable gross weight. The left crosswind is perfect for a tail dragger because it's coming uphill. If we had a crosswind coming down off the hill we would have to reject the strip as unsafe.

The aim point is now at about twenty five degrees aft of the wing-tip, so we roll into a bank, turning to base leg and reducing the power a little to allow the airspeed to slowly decrease to the approach speed of seventy mph. As we get on approach speed, I crank the trim handle to trim out any required pressures for elevator control. We'll be turning to final very shortly so we do a quick check for big birds and airplanes in the final approach lane. With the over-shooting crosswind, I start the turn to final early, so as to avoid over-shooting.

After the turn to final, the aim point slides down almost to where it should be, as I pitch to the glide slope by putting the surface aim point where it ought to be above the engine cowl (based on experience). Simultaneously, I reduce power a little to hold the airspeed. I take a quick glance at the airspeed indicator to verify our speed and adjust the throttle a tad, as necessary. My sight picture of horizon, engine cowl, and aim point, the experience-based product of glide slope and angle of attack, looks good. I do a careful scan of the strip from approach end to the far end, checking the sides of the area as well as the middle. There's nothing new; it still looks "good-to-go." Things are stabilized, so it's one last refinement of elevator

trim to adjust for the slight reduction in power, and then we'll leave the trim where it is.

From now on it's “eyes out of the cockpit.” At about 200 feet AGL, I take a careful look around the perimeter of the landing area, checking for obstacles, moving or stationary.

Quickly, and by feel, I check the position of my boot heels with respect to the brake heel pedals and check the firmness of the brakes. Then I release the brakes while keeping my heels in position. I remind myself that any braking will <u>not</u> involve simultaneous application of both brakes. (Single side application of brakes will be only quick jabs, and then only if rudder is not enough to stop any yaw which could be created by rolling over a small, soft patch or such).

Now we're descending close-in toward the aim point and before it goes under the nose, I reduce the power a little, relying on the change in engine sound. Then I start a flare at a point based on experience in this model of airplane while trying to be as smooth as I can. As the wheels reach a few inches above the runway, I use slowly increasing up elevator to hold it there. Because we are slower than the trimmed airspeed, I'm holding a positive up-elevator (back stick) pressure. This is important because there is no "dead band" in the elevator control that <u>would</u> exist if we were “trimmed-out” for the slower touchdown airspeed. Therefore, the elevator will be immediately responsive to even small changes in back pressure. This is very

important to any pilot striving to be precise. Little things add up when trying to do your best.

The aircraft touches down only on its main wheels, slightly on the back sides, but not too close to a three-point attitude. Keeping my eyes forward and out of the cockpit, I immediately but smoothly ease the stick forward to hold the tail up, and then retract the flaps. Raising the flaps makes the rudders more effective and also ensures that a sudden gust is less likely to put us back into the air. Holding the tail up reduces angle of attack and puts more weight on the main gear, protects the tail wheel and gives you better visibility over the nose in the early phase of roll-out.

As the airplane slows, the tail wheel settles on the surface and we “drift” left (downhill) so as to have room to make a right one-eighty when the speed is right. By aggressive use of right rudder and prop blast, when slow enough, I'll swing the tail through the downhill direction in order to give the C.G. lots of momentum for making the one-eighty turn on the tilted surface. As I do this I smoothly go to forward stick to lighten the load on the tail wheel and to reduce the twisting moment on the tail supporting structure. From experience, I have learned to deflect the elevator in coordination with the amount of power I’m using to make the turn in order to keep the tail wheel lightly on the surface. You will also learn from experience. To the uninitiated, the blasts of power will seem rather aggressive. Ailerons and elevators are also used to counter wind effects as I taxi downwind to the open

area on the uphill side of the strip. We stop and shut down. Then we exit and push the aircraft off the strip to our tie-down site. Then it is: turn the fuel selector "off" before we unload, tie down with wings level, and secure the airplane.

Now, it's time to walk up and down the strip to find things I <u>failed</u> to see in the pre-landing survey. (No one is too experienced to learn more!) Also, I'll check our wheel tracks to see if the airplane touched down where I had intended. This post-flight chore is always educational and sometimes humbling.

CHAPTER SEVEN

BEACHING OR PARKING, SECURING YOUR AIRPLANE, AND TAKING OFF

Securing the Seaplane

With a seaplane, your planning for a saltwater beach must be geared around the effects of tide and wind. As you taxi to the beach after landing, line-up a near object with another object more distant (i.e., make a "range"). Now you can readily detect the direction and strength of any tidal cross-current and/or cross-wind simply by keeping the plane on your makeshift range. Then, knowing the wind strength/direction by its effects on water surface and the airplane, you can deduce tidal strength and direction by observing what must be done to remain on your chosen range. Doing this as a habit will help you remain aware of your situation with respect to wind and current.

Shut down and temporarily secure the airplane, allowing for what the tide is doing <u>now</u>. After getting the plane stabilized, run a quick check of structural integrity. Just stand on the float that's most water-borne and spring your weight up and down on your forefeet. If you hear unusual noises, check brace wires for tautness as well as N-struts and bottoms for security. After that, take a walk on the beach.

Verify the underwater rocks you spotted from the air. Also, find the ones you didn't see. The idea is

to continually improve your airborne evaluations.

When positioning your floatplane overnight on a salt water beach, plan for the highest and lowest tides expected during your stay. If the beach is exposed to wave action, the airplane must be dragged up and secured beyond the highest tide, if possible. Of course, we're talking about small, unloaded floatplanes. Using strips of plywood under the keels will help on this chore.

Alternatively, the seaplane can be moored to a buoy or to a “siwash” anchorage so as to keep it from touching bottom at tidal low. (See figure 7-1.)

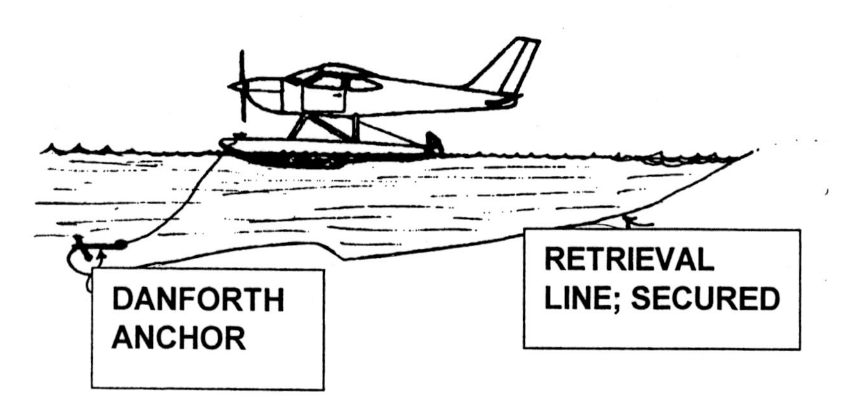

SIIWASH ANCHORAGE
FIGURE 7-1

Allow for higher than normal tides when the wind is expected to be onshore. This is not a good alternative when the winds are expected to be onshore above about five knots, or a little higher offshore. In these cases, if you can’t find a sheltered anchorage, consider other options.

With an amphibious seaplane, check the firmness and gradient of the beach before taxiing onto it with wheels down. The beach must be solidly firm and very shallow. Use thick (3/4 inch) plywood strips (two for each side) under the bow wheel(s) as a precaution to keep them from sinking in and being damaged. Otherwise, it's easy to damage the nose wheel structures on amphibious floatplanes. Some two-by-sixes (two for each side) will help under the main wheels. Careful avoidance of soft spots saves a world of grunting and groaning, to say nothing of avoiding an overheated engine.

For any stay beyond one tidal cycle, if you can work the plane in behind an inlet point of a nearby lagoon, you can expect better protection from wave action. However, in many cases, this location could expose your plane to sand blasting by strong wind coming over the outer strand.

One practical solution for extended stays is to build a log raft anchored with heavy lines in a protected bay or lagoon. With some manipulation of raft tilt and buoyancy with heavy weights (rocks), you'll be able to drag or power the floatplane onto the raft. Point it into the wind, securing it firmly to the raft. Rig soft spoilers on the top surfaces of the wings for possible high winds. (If high winds are actually in the forecast, it's a better idea to head for home while you can.)

Right about now, you'll be appreciating the benefits of basing a seaplane in fresh water. For that you can rig a "siwash" anchorage, or a log

ramp on the beach.

If temperatures are near or below water freezing temperature, and with all the airborne moisture near bodies of water, your engine controls could easily freeze up, even around salt water. You can avoid problems of starting later by setting the mixture and throttle in their start-engine positions after you complete the shut-down. As a safety measure, disconnect the battery ground until you're ready to restart the engine (after de-icing your airfoils). As the engine warms up, the mixture and throttle will be freed.

Do not secure a seaplane overnight in a water-borne situation if you haven't first checked its water-tight compartments for leaks. You should, of course, never fly a seaplane without a float/hull repair kit. Also, never leave a seaplane water-borne if there is a possibility of snow. Heavy snow on tail and floats will start the rear of the floats down, to the point that water will start entering through non-water-tight top panels; and the sinking begins. In sum, leaks that have not been sealed and/or snow accumulation not cleared can sink the float plane.

Taxiing and Securing the Land Plane

If you have landed your wheel plane in soft terrain and with wind above about ten knots, be careful to avoid the typical tail dragger "nose-up trap." What happens is that the pilot will experience difficulty turning the airplane out of the wind. Consequently, instead of shutting down the engine

and manually swinging the airplane around, he will use wheel braking to assist the turn. Then, as the wheel being braked builds its own "chock" in the soft turf, he adds even more power. Bingo, the tail comes up quickly, possibly assisted by tailwind if he's turned far enough, and the nose drops to a prop strike with the surface. This even happens to experienced pilots.

The above sequence will happen even quicker if the plane is not heavily loaded. If you must use power in the magneto-check region (1600-1800 rpm) or higher to turn out of the wind, beware of the "nose-up trap." Never ask a passenger to swing the tail in this situation, unless it's absolutely necessary. Do that only after you've given that person adequate on-the-spot training.

Another way to avoid the nose-up trap is get some taxi speed before your turn downwind and use rudder/elevator/power combinations to lighten the tail in the first part of the turn while using rudder against prop blast to start the turn and keep it going. Then follow with appropriate aileron to counter wind effects, as well as neutral or down elevator as the tail gets up wind. This takes a fairly wide area. It also takes some practice in benign wind before you try it in strong, gusty winds. Take care here.

Try to point the plane into the wind prior to shutdown on loose surfaces. This reduces the possibility that you'll have to start the engine in a strong tail wind and then, to get moving, blast small

gravel which can be blown back into your prop by the wind.

Before shutdown, with the engine at idle, release the brakes and see if the airplane will stay where it is. If not, put the plane in a safer, more level spot before shutting down. Then set the brakes and shut down. If you are in a ski plane, try to find some brush tops above the snow to get the skis up on, to increase surface friction as well as to avoid freeze-down. In precipitous terrain, it's very important to verify that the airplane will stay put before you shutdown.

Also, prior to shutting down your airplane, determine how you will get the wings level to prevent fuel loss overboard. And you'll want to select "Off" with the fuel selector after shutdown to prevent wing-to-wing transfer, followed by overflow through the low tank vent.

Before you shut down on a beach, look at the wheels. If you and the plane are sinking, taxi to high/dry ground. Or, better yet, takeoff and don't land there again!

If you shut down and then notice the sinking, you might have to get out your little survival shovel and quickly scoop out a shallow ramp. Don't forget to make room for the propeller. In any case, you must act quickly to save your airplane. Fortunately, the author has had to do this only once. After quickly scooping sand and letting some air out of the tires, we left, never to land on that beach again!

After shutdown, do a post-flight inspection of the aircraft. For the wheel plane, this means chocking the wheels and taking a thorough look at the landing gear and supporting structures. This also includes the struts and flying (brace) wires that project below the wing and tail surfaces and can be damaged by unnoticed low obstacles.

For the ski-plane, you'll want to look at the undercarriage, rigging bungees, and cables. You should ensure that the skis are on brush, logs, or a couple of short two-by-fours. Do your best to avoid meltdown of friction-heated skis into the snow, followed by freezing. If there is any slope to the surface, parking with the nose pointing about forty five degrees to straight downhill works well on gentle slopes. The center of gravity bears more on the downhill ski and causes it to dig in more. Make sure it's holding steady at idle before you shut down. Then, you'll want to anchor some tie-downs to stabilize things.

If you have parked the aircraft in the wilderness, do not leave food or food debris of any kind in the airplane. Carrying fresh fish that's not triple-bagged is really asking for it. The bears and other critters will find food or fresh game/fish regardless, and they will damage your airplane trying to get to it. In the process, they'll sometimes bang up your fuselage and control surfaces. Eating or storing food in your airplane is in the same category as eating or storing it in your tent -- it invites four-legged trouble.

In tying down your aircraft, avoid the use of rigid devices such as chains. Use ropes or straps. You want the tie-downs to stretch and flex, otherwise something in the airplane could break with the stresses from buffeting winds. With the airplane headed into strong wind, its normal tendency without spoilers on top of the wing will be to fly.

If the elevator control can be secured in the forward position and the plane allowed by the tail tie-down to come to a slightly nose-down flight position, the airplane will be able to handle extremely high winds from the front. With the elevator control secured in the aft position and a tight tail wheel tie down (the traditional way), you are risking a flip-over.

For example, an acquaintance told me about parking at a field in Fairbanks and leaving his small tail dragger with the stick secured forward, and only chocks and single tie-down ropes on the wings plus a long rope on the tail wheel. The airplane luckily was pointed in the right direction. During that night there were freakish high winds. The next morning, he found his airplane in pristine condition and several trashed, flipped-over tail draggers around it. They had all been pointed the same direction the day before. The only differences were in elevator positions and the longer tail wheel rope. That's convincing.

If you did it right, your takeoff was planned before you landed. Logically, you should ask yourself whether the conditions (wind, tide, current, ceiling, visibility) have changed since you did that initial planning. If so, consider how those changes will affect your takeoff and departure.

Here's where the professional pilot has an advantage and will, most of the time, leave the landing site within the hour after unloading/loading. There likely will be little change to ponder. But the owner-pilot could be parked in the wilderness for a couple of weeks, with many changes taking place.

For the land plane, a pre-takeoff walk along the takeoff run is helpful, to check for obstacles, verify location of reference go/no-go points as well as to refine the departure route. It is also essential for picking out good takeoff directional reference points in a dogleg takeoff run. Or, you can make your own reference points. For example, if you want to make a go/no-go reference, or perhaps establish a point at which to start a dogleg turn, you can make a rock cairn alongside the takeoff run where you can easily see it. If there are no rocks available, hang a piece of bright flagging on nearby brush. This helps to reduce the possibility of confusion.

In an airplane requiring the pilot to reach low in the cockpit to initiate the lowering of flaps, it is wise to set the flaps at least to the first notch to avoid

limiting your view after starting the takeoff run. Then, before engine start, practice lowering the flaps as required while keeping your eyes forward, out of the cockpit.

If you plan a dogleg takeoff on skis, allow for wider turns than that which is possible on wheels. Select “both” fuel tanks or the one on the inside of the turn to avoid unporting a tank. Have a sure-fire reference point where you will be starting the dogleg turn, and don't be too near lift-off airspeed at the turn. Keep the wings level with ailerons and don’t lift off during the turn, reducing throttle as required. Then, <u>after</u> the turn, go to takeoff power and lift off with level wings.

With a sidewise-tilted runway area or beach, to avoid fuel tank unporting, take care to select both tanks or, if "both" is not available, select the upslope tank provided there is adequate fuel in that tank. Start your takeoff roll with the nose of your <u>tail dragger</u> pointed about twenty or so degrees <u>down-hill</u>. By the time you gain positive directional control you should be parallel to the "center-line." If this doesn't work, shut the engine down and regain control of the airplane with rudders and, if necessary, alternating braking (quick jabs only) and full up-elevator. Take a bigger downhill angle on the next try. Better yet, reduce the load to move the C.G. further forward and shuttle smaller loads out to a consolidation point.

Loss of directional control on a side-slope can be simply a brief sporting moment if you get the

throttle off, the stick/yoke back, the engine shut down, and continue doing your best to regain directional control. The old axiom you learned prior to your first solo still works: you are better off being out of control with the propeller stopped than being out of control with full power. (More than once on a steep beach, the author has had to say on the intercom: "Brace yourself, we have to abort this take-off.") No damage was ever done.

For the seaplane, you should always pump out water accumulated in the floats prior to start. The amphibious floatplane that uses hydraulics in its wheel raising/lowering system will reveal in-float hydraulic leaks with red fluid as the pilot completes pumping out the water. Hydraulic fluid will float on top of the water inside the floats and you won't see it until the last few strokes on the pump.

Once, at Bear Lake on the Alaska Peninsula, after a stop at the Fish and Game fish weir, I pumped the floats on our amphibious DeHavilland Beaver and found a small amount of red fluid in one of the floats. This appeared to come from a small, new leak. After engine start, the system came up to and held steady, normal pressure.

But during the latter portion of the water takeoff run, I heard the electro-hydraulic pump start cycling on and off, indicating the system had lost its integrity because the leak had gotten bigger. Because gear-up was selected, the leak had to be in the up-line part of the system.

Having already thought about what this would mean, I immediately used rough water technique to get airborne. (The cycling pump was doing its best to keep the pressure up and, in the process, pumping fluid to the leak.) Then, I quickly turned off the pump. The main landing gear then came down due to the rupture in the up-line. We prepared for a wheels-down landing at the Port Moller gravel strip, using the system's down-circuit and remaining fluid to lower all gear.

None of the above took skill. The main point is that finding a bit of fluid in one the floats prior to start had given me pause to think about what I'd do if pressure were to fail during takeoff, and how I would know that a failure had occurred. Since we were near lift-off speed (the deciding factor) and the main gear is held up by hydraulic pressure, the obvious solution was to get that Beaver airborne quickly.

There was no particular urgency to depart because I had tools in the rear baggage compartment and spare hydraulic parts in the floats. I could have just as easily repaired the line at Bear Lake as anywhere else. (And I should have!) Being near liftoff speed when the leak became serious, I had to get that Beaver in the air before the main gear lowered by gravity and put us on our back in the water. Thinking about it in advance provided an advantage, despite my poor decision not to repair on-site before takeoff.

As you warm your engine, think about the requirements for that particular departure. Visualize what you'll do with the airplane after lift-off, and how the wind combined with obstacles may influence your plan. Above all, don't be automatic. Your departure can sometimes be more exciting than your arrival, due to weather and/or terrain clearance requirements.

For example, a pilot may have had an easy time with a low-power, circling descent into a valley, followed by a routine landing at a friendly off-airport site. But on departure, he gave no thought about his usual climb-out going through a narrow canyon and getting above the terrain at the far end with no problem. But now he has a heavy load and is getting a climb-flattening tail wind – hopefully, he will recognize that he has a serious problem in time to do something about it. This kind of situation is common in mountainous terrain.

For early recognition of climb-out terrain clearance problems, after you establish your climb course, pick out the highest terrain ahead, note its position relative to a reference point on your cowling. Then, as you climb, frequently note whether or not its relative position is dropping below your cowling reference point. If it's not, or you're not sure, turn and modify your departure. There is always the chance that, as you climb, the wind will shear to your tail. So, do not assume that the climb-phase wind will remain constant in

direction or intensity. It is much more likely to change than not, especially in the mountains. Check often in the climb.

If you are climbing into a head wind your angle of climb will benefit. However, be ready to turn around at the first encounter with the likely down drafts if there is high terrain ahead. In that situation, it will help to approach the high terrain at a forty-five degree angle, thus making it easier and quicker to turn away and climb higher before attempting further progress along your route.

CHAPTER EIGHT

WILDERNESS SURVEY PILOT PROCEDURES AND TECHNIQUES

The purpose of this chapter is present knowledge that, although directly related to rather specialized flying, can be useful in making any pilot safer in wilderness flying. A pilot who can perform basic bush piloting tasks safely can be an effective wilderness survey pilot, and vice versa. Some of the techniques presented in this chapter can be useful in non-commercial, recreational flying and in the process make that flying safer.

Wilderness Survey Flight Crew Coordination and Training

Let's assume you are a Commercial Pilot assigned to carry out a wilderness survey contract for the impending season. You, as pilot in command, are responsible for your crew's ability to work together to achieve each flight's objective. Therefore, you must understand the capabilities, limitations, and power requirements of any technical survey equipment installed or proposed to be installed inside or outside of the airplane. External antennas must be clear of the propeller arc, flight control surfaces, air(speed) brakes, landing gear and Its retraction and extension arcs, and wing flaps operated to full extension. If feasible, antennas should not obstruct the pilot's and crew members' views. They must be mounted securely and, if possible, beyond the propeller blast

to minimize vibration and possible damage.

If equipment power requirements are high, expect that you could have a cooling problem and/or a fire hazard without careful installation – no matter how temporary that may be.

Even though the crew member(s) (wildlife technician, biologist, etc.) might have more flight time than you, as pilot-in-command you are responsible for training them in <u>your</u> in-flight procedures. On the other hand, once you find out the particular requirements of the crew with which you will work, you may have to modify your own procedures for reasons of safety and efficiency. You must be flexible, yet retain a strong eye for safety.

It is highly recommended that you schedule a preliminary introductory session with your crew at the airplane and with full flight gear on or available. Waiting until just prior to the first scheduled departure does not "fill the bill." You could be in a hurry and the training would suffer.

Getting Started With Your Survey Crewmembers

During the preliminary session you should cover the basics:

<u>Use of seat belts and shoulder harness</u>. Always cinch the seat belt snugly prior to taking slack out of shoulder straps. If the seat belt is loose, with a sudden stop, the occupant of that seat will slip

down and forward without protection. Keep belts and harness on at all times - get the pilot's prior permission if unbuckling is necessary. In an emergency the pilot will state when it is allowable to unbuckle. The importance of looking at the seat belt buckle when unbuckling should be stressed. The flip-over should be discussed, including the hazard of falling on one's head upon release of buckle. Falling on your head can be prevented by bracing with one hand on the cabin ceiling before you release the buckle.

Cabin egress sequence and procedures, upright and inverted. This must also be covered in detail. Make assignments for taking survival gear during egress from airplane. Run a practice drill. Cover different contingencies and how they logically would change the egress sequence or procedures (for example, if the airplane comes to a stop with an engine fire on the right side, or if the pilot is incapacitated, etc.) Emphasize that even when inverted, things (like doors) that were on your right when erect are still on your right when inverted. Remembering this reduces somewhat the confusion that comes simply from being upside down.

Survival equipment and supplies, airplane and personal. Point out that each person must have an on-person survival kit. This, because there can be many situations where you will not be able to take your major survival equipment and supplies from the airplane before it is lost in fire or sinking. And thus, what you're wearing is what you're going to have for survival. A sport fishing vest with pockets

works well for this.

The airplane survival kit and supplies should be opened and examined in detail. The crew should be invited to comment on how the kit and supplies can be improved. (Field technicians and biologists almost always are highly experienced in the outdoors and can provide excellent suggestions for you to improve your own camping/survival gear.) The advantages of woolen flight clothing and other fire-retardant fabrics in the fire situation as well as in long-term survival should be discussed. The use of helmets should be encouraged.

Lookout doctrine. The crew should assist when able to look for other aircraft and should report their sightings in the clock code, amplified by the words "level," "low," or "high." To minimize unnecessary reports, the pilot should report his own sightings to the crew - either on intercom or simply by pointing.

Motion sickness. Find out to what degree, if any, crew members are prone to motion sickness. Ensure that the aircraft is equipped with sick sacks. Brief the crew on methods of fighting motion sickness: keeping eyes on, or making frequent looks at, the distant horizon, limiting amount of time with eyes inside the cockpit, breathing plenty of fresh air, strapping-in securely, pilot techniques to limit rates of pitch, roll, and load factor.

Mission objectives and procedures. This subject should be covered prior to every flight. Crew members should include their own briefing of the

pilot on the equipment to be used, the basic search or survey procedures and altitudes to be used, maneuvering limitations, and the ways in which the pilot can assist in achieving the overall objectives. The pilot should cover current and forecast weather.

Flight safety priority. The pilot should make it clear that the first priority on all flights will be safety of flight. Stress importance of cabin discipline, especially during takeoff and landing. Give examples. Invite questions.

Basic Radio Direction Search/Survey

There are wilderness survey flights that include visual and/or technology-based low-level pipeline security and prescribed flight paths and or grid search patterns using technological sensors for mineral deposits. Much of this book's information can be applicable to those pursuits. But the focus here is on techniques associated with search and survey of fish, mammals and birds.

There are continuing leaps in fish and wildlife radio search and survey technology, including space-based systems. Yet, there remains for many organizations, the use of basic, folded dipole, passive receiver antennas. The transmitter is strapped onto or implanted in the animal. Each transmitter in a given geographic area operates on its own separate (discrete) radio frequency.

The basic airplane-mounted system uses two

receiver antennas, one mounted under each wing of a high wing airplane. The right antenna receives a stronger signal when the transmitter is to the right of the airplane. Likewise, signals from the left are better received on the left antenna. The signals are received on a line-of-sight basis, but they can be reflected and distorted by rough terrain. Generally, the higher the receiver altitude, the greater the distance at which the signal can be received.

If the airplane is headed directly towards (or away from) the signal source, the operator can verify this by switching from one antenna to the other and receive equal signal strength. If the signal fades, then the plane must reverse course to close the signal source. When getting closer to the transmitter, the signal volume usually will start to build on one side or the other, indicating that the transmitter is a little to the right or left. At the moment of maximum signal to one side, the plane is pointed ninety degrees to the signal source. By virtue of the sequence and geometry of events, you are always past the transmitter before that fact is verified unless, of course, the crew gets an early visual contact.

Throughout the search evolution, the biologist or technician will be switching from one antenna to the other, comparing signal strengths, while coaching the pilot onto an intercept course. There are other methods of mounting the antennas. Different mounting will change your required localization techniques. The best method to be used in a particular area will normally depend on terrain

roughness. Generally, the higher you climb, the greater chance you'll have of picking up the initial signal.

Flight Techniques for Search and Survey

At the terminus of each individual survey, the biologist or technician is interested in the identity and condition of the animal, exactly where it is, and what it is doing. This means you'll be flying at low altitude. And it calls for special rules to avoid getting talked into a collision with the terrain. The following recommended rules have worked well for the author over thousands of hours flying visual and radio search/survey flights.

Fly at an airspeed between Vy and operational Va for your gross weight. This keeps you and your airplane ready for full power climbs if required to overcome a down draft.

There will be times when you'll be asked to fly low and slow for condition/quality surveys of animals. Before you slow down and lower maximum performance flaps (usually twenty degrees), make a safety check. Ensure that there is plenty of maneuvering room and that the area is not contaminated by down drafts that can put you down, or by snow showers that can give you white-out.

In really frigid weather (below -15 degrees F.), do an S-turn, look back, and make sure you're not

creating ice fog. It can fill up the area and also create white-out. Therefore, you'll want to work starting on the downwind side in order to accomplish anything. If things on the surface, including the animals, look fuzzy, you could be looking at near-surface white-out conditions. Avoid getting low.

Plan your low-slow fly-by into the wind, in order to minimize ground speed and maximize look-time. Fly no slower than 1.3 Vs0 plus gust factor. Trying to do a pylon turn rather than a race track pattern will get you slow, low and steeply banked. That's not good for longevity and, besides, it doesn't give the biologist a good look-time, to say nothing of what it might do to his or her breakfast.

At the upwind end of the race track, after going by the object of inspection, speed up a little, climb a little and then turn to the "downwind" leg. Make this turn towards the downhill side. Stay ahead of the airplane but don't rush; be deliberate and methodical. Be flexible but thoughtful: without impulsive changes in your immediate plan.

Avoid flying uphill on the inspection leg. If this means flying cross-wind on the inspection leg, that's okay. If you are cross-wind on the inspection leg, put the biologist's side on the downwind side if you're in a side-by-side cabin. Your crab into the wind will give the biologist an even longer look than without cross-wind. Pay attention to giving the biologist the best available lighting with respect to sun direction. Often, this will make a real difference.

Getting lower than one hundred feet tends to limit the biologist's look-time and counteract the benefits of slowing down. Nevertheless, you could be asked to get even closer. Don't do it unless you have perfect conditions.

If you ever get trapped by white-out while doing low-slow animal quality/condition surveys, don't ignore the possibility of making an immediate landing alongside the animals -- they could be your only vertical reference outside the airplane. I've had to do this only once, when a passing snow shower surprised us. It worked out well despite the bumpy landing and objections from the rear seat about the short notice. But it sure as heck beat flying into a hill while groping around climbing out of that valley.

Fly gently. Roll in and out of banked turns in smooth fashion. To do this, you need to train the biologist to let you know, ahead of time, when and in which direction he next wants to turn. Try to return the favor if you are following a creek and you are the one deciding when the turns will start. With a little notice both ways, you'll develop smooth crew coordination and two-way respect.

Use forty-five degrees bank as your normal maximum for turns, but favor thirty degrees of bank in practice. This gives you some leeway for safety's sake in tight spots. Once, I rapidly rolled into a sixty degrees banked turn over an archeological site we had been looking for while doing wildlife surveys. The site was level and really hard to see from anywhere but directly overhead. By pure

chance, we flew directly over the site. Needing to get some visual reference markers immediately, I quickly cranked the Super Cub into the steeply banked turn, trying to be smooth. The biologist, who had never gotten sick in my plane during our many flights together, promptly upchucked. Despite my apology, he never flew a contract flight with me again. (As we used to say: “that’s the breaks of naval air.”)

When following linear features, such as salmon streams on fish surveys or following creeks and valleys when looking for cruising carnivores, fly downhill. With the least complication, flying uphill can put you in a world-of-hurt for altitude and/or maneuvering space quicker than you would imagine. Once I was making a "small" exception to this rule, when the muffler on our heavily loaded Super Cub started to self-destruct. At our standard 175 feet AGL for fish counting, we were at the upper reaches of a short salmon stream fork, tightly enclosed by terrain. My plan had been to go uphill on that short fork, then swing over a ridge and come downhill on a much longer fork, saving time and fuel.

Suddenly, the engine stumbled and started winding down. Carburetor heat was put on, mixture enriched, fuel tanks changed, and I started a one-eighty towards downstream. At the same time, suspecting muffler pieces clogging the exhaust, I slowly reduced throttle setting and started getting power at about 1950 rpm. (It turned out to be the muffler.)

The tide was at spring high; planned that way in order to spot the fish schooling close-in at the mouth of the stream. However, high tide also ensured there would be no suitable beach for landing. We made it into a small strip on a bay about five miles away, by the "skin of our teeth." Starting right then, there would be no exceptions, however small they may be, to the “downhill rule.”

Give yourself plenty of room for turns. For example, if you're in tight quarters and planning a left turn, first move over as far to the right as you safely can to maximize the lateral space available for the turn. When there's no room for error, it makes good sense to create whatever extra room you can.

In planning your room-for-turns, do not count on using flaps. Then, you will have available the maximum performance flap setting whenever you see that a turn is going to take more space than you expected. In any case, if due to the higher induced drag of a turn, the plane slows to near approach speeds, it will ordinarily be advantageous to lower flaps to the maximum performance setting (usually 20-25 degrees) while adding power.

Specialized Search and Survey Techniques

Fish Radio Surveys. As soon as you determine which of your antennas is the most sensitive, plan to use that side for reception. Transmitters implanted in fish are difficult to receive, even in the

shallow water of a creek. Accordingly, you'll likely have to fly below 200 feet AGL with speed appropriate to the amount of maneuvering required to stay close aboard the creek. Because of the low altitude, this is sporty, very high workload flying, and you must anticipate required maneuvers.

In other applications, you might be doing a shoreline search or a grid pattern over known anadromous fish routes. This type of survey is used primarily with steelheads that have been implanted during a previous trip to fresh water, or with salmon implanted at sea while migrating.

Visual Fish Stream Surveys. In order to make management decisions regarding salmon fisheries, the area management biologist needs to know, among other things, how many salmon (by species) have escaped fishing and made it into their respective spawning areas. A certain target percentage of the predicted total salmon run to any particular stream system must have escaped into the system before fishing restraints are eased. Fish are counted on foot, by sonar, at weirs, and by air.

Visual airborne salmon stream surveys are ordinarily flown going downstream at about 175 to 200 feet AGL, staying just clear to one side of the stream, and following the stream closely in order to avoid missing any fish. The biologist will determine the desired altitude and lateral position.

Many salmon streams have sharp curves that are too tight to be followed safely by airplane. Instead,

you have to make 270-degree turns away from the curve, after a slight delay, and then rejoin the stream bank at the curve where you left it. In many curves, where vegetation is overhanging the outside cut-bank, there is created a good temporary refuge for salmon going to the spawning grounds. But these fish are hard to see. So, you must fly to the inside of the curve when coming out of the two-seventy. (See figure 8-1.)

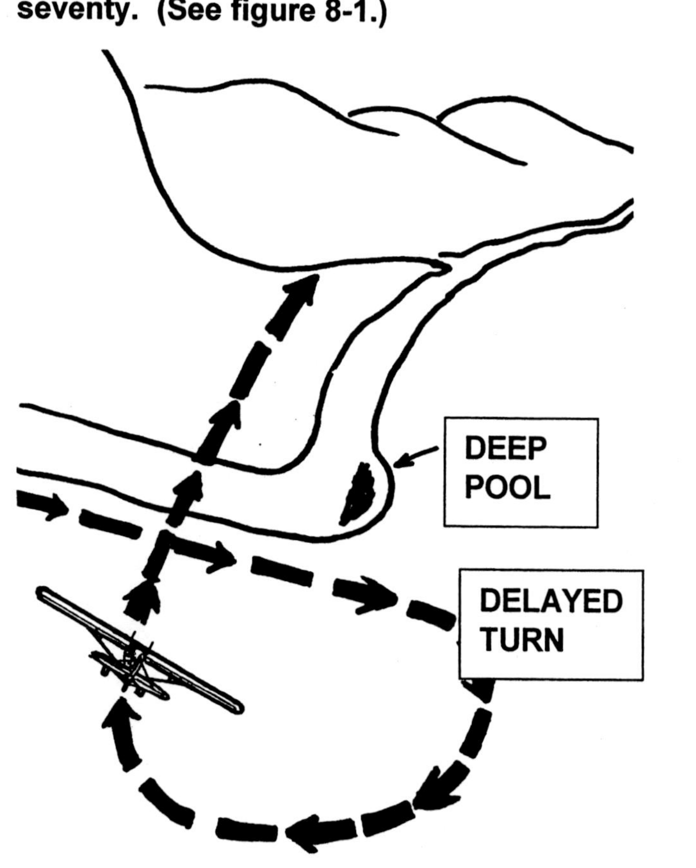

OUTSIDE 270-DEGREE TURN
FIGURE 8-1

At many locales, the stream might be so "snakey" that you have to be creative to give the biologist an ordered look-see. Be careful here, you might wind up going back up stream because the downstream flow visual cues are not so apparent in lazy, slow running streams. After several fancy turns and reversals it's easy to do. I hate to admit it, but I've done that, much to my dismay -- and to that of the guy in the rear seat. (Just goes to show: not all of us so-called "real pros" are very "real"!)

Low Altitude Surveys and Flight Safety

Work your way into low altitude surveys with cautious alertness. Beware of becoming relaxed and over-confident. It is not the safest of pursuits. Do only that which you know to be safe, don't fly too aggressively (be smooth), know and recognize pre-stall buffet onset for every individual airplane you fly. Each individual airplane has distinctive little aerodynamic quirks – noises and flight control feedback -- that give early stall warning when pulling smoothly just below the pronounced accelerated stall buffet. Caution: Airplanes with STOL modifications are understandably less likely to give significant pre-stall warning, due to laminar flow and other stall delay "improvements."

Never fly low altitude surveys when over the normal maximum allowable gross weight (MAGW) -- even if you do have a special FAA waiver. We had a waiver in Fish and Game, but we used it only for cruising with extra belly tank fuel. We could fly to the far end of a long survey route, emptying the

belly tank in the process. Then we would start our low altitude work at MAGW or below. Anytime you load a small airplane close to or beyond MAGW, its center-of-gravity usually will be near the aft C.G. limit. Do not use more flaps than you really need, because your nose-down pitch capability could be drastically altered when extending flaps.

In sum, being too heavy not only cuts performance but could also limit your ability to control the pitch of the aircraft when slow and with flaps down. So, with a waiver, in that condition you sensibly should only cruise while making a consumable (gasoline) the only permitted weight above MAGW, and use that consumable to get down to maneuvering weight before landing or beginning any maneuvers. We're talking serious stuff here; things that can put you out of control and down in a heart beat if not dealt with very conservatively.

As mentioned before, there are small nuances of difference between airplanes of the same make and model. So, don't assume anything about any particular airplane when you're flying that airplane "on the edge."

Never fly in weather that you are concerned about. When you're working at 175-200 feet AGL, the weather can come down on you undetected if you don't stay alert to its trends. If you get concerned, it's past time to go home or, at the very least, climb up to better check the ceiling and visibility. Nothing is important enough to justify infringement on safety.

The Importance of Fitness

Salmon stream and other types of survey are long periods of low altitude, precise, and physically demanding flight. The pilot must stay in good physical condition and be rested to go day after day, flying eight to twelve hours a day of stream surveys. Being in my fifties and sixties when committing "survey madness," I realized that my fitness could "go down the tubes" if I didn't work at it. So, regardless of how tired I was at the end of a long day, I did a daily four-to-five mile run (while carrying a rifle for bear protection), plus weight lifting and calisthenics to stay fit. It's time-consuming but worth it. Moreover, I didn't need nearly as much sleep to be well rested when pursuing fitness with a passion. In fact, you always come out ahead in more ways than one when you make fitness an essential ingredient of your regime.

Bear Collaring

Both fixed and rotary wing aircraft are used in teams to collar bears for tracking and study. Since the females are critical in the rearing of new generations, they are more likely to be chosen for collaring. Females with cubs give more return for the effort in terms of number of bears studied. The biologist who chooses the bears to be collared usually rides in the fixed wing aircraft. The airplane pilot stays well above the bear at about 1000 feet AGL, turning as necessary to follow, while the biologist calls in the helicopter with the darting, collaring and tagging crew. Ordinarily, the job is

easier if the bears can be selected as they descend from their dens along the snowfields but not yet at the lower vegetated areas.

After the bear is darted with tranquilizer, the helicopter moves away. Apparently, turbine noise terrifies bears and can cause them to run even faster and to dangerously overheat before succumbing to the drug. The airplane pilot maintains visual contact on the bear in its mad dash before being overcome by the drug. If the bear makes it to vegetation, you as the pilot will be challenged to keep it in sight.

If it gets into dense brush, you'll simply have to look for shaking and disintegrating vegetation as the bear plows through it. If the bear gets on a trail with overhung branches, it might be harder to maintain a visual, but don't give up and you'll spot it eventually.

Once the bear is down, the helicopter is called in and coached as needed to the site by you. Then the helicopter delivers the collaring/tagging crew, stands by, and should keep its radio alive. You should remain high and watch for movement by the darted bear, as well as for any other wildlife coming on scene. You should pass such vital information to the helicopter.

When finished, the collaring/tagging crew gives an antidote to the darted bear and promptly boards the ready-to-go helicopter. (First-year cubs are not tranquilized. Rather, they are caught and ear-

tagged. Strong little critters, they come equipped with sharp teeth and claws, thus providing sporting exercise for the robust young men involved.)

Then, off you go, looking for another candidate. Things don't always go as planned. Consequently, you have to be flexible and ready to pass information to the helicopter if it will be of assistance.

Springtime First-year Cub Survey

This is a visual, new-family, bear population survey. Sows with first-year cubs will normally stay high in the hills until food is plentiful at lower levels. The nursing sow feeds, among other things, on high-nutrition sedge grasses growing along the edges of receding snow patches. In this survey, the pilot's job will be to fly along altitude contour levels helping the biologist spot bear families. Obviously the levels will be where there are snow patches, not too high and not too low. The right levels will vary between the sun-warmed, south sides and the shady, cooler, north sides.

As you fly along close to the steep terrain, you'll have to be alert to avoid traps presented by terrain changes. Don't do this type of survey when it's windy or when precipitation is falling -- the bears will be "holed up" out of sight anyway. You need light winds and smooth air for safety.

Mountain Goat Survey

Looking for mountain goats is best done when mountain snows have melted as much as they're going to, in summer, and yet before the first August snows have fallen in the high country. The biologist will usually have a good idea of the altitude contour level that should be flown. You need sunny weather with <u>calm</u> winds because you are going to get "cozy" with some very rugged, vertical terrain.

Flying close to the vertical goat habitat, you should scan well forward to anticipate required turns and avoid dead-end canyons. At the same time, you'll be looking just where you should to have the best chance of seeing goats.

Report your sightings to the biologist; but it is what he or she sees that counts (as with any animal survey). Early reports are essential. The goats will have a yellowish-white color compared to white snow patches. As you get closer to where you first have seen goats, they will have disappeared, seemingly into blank rock. Look higher.

Bird Radio Survey

You may have occasion to conduct survey of large birds having radios mounted in "mini-backpacks." You should conduct the localization of the bird from 1000-1500 feet AGL. If you try it from low levels, you'll be scaring the bird from place to place. Thus, it will become a very frustrating exercise. (Been there...done that!)

After localizing the bird, descend and identify it quickly before it flies away. It's usually the young birds (yes, the camouflaged ones) that you're tracking. Just make sure you have a good fix before you descend, and try to spot the bird as quickly as possible, before or as it flies away. Look for movement; it may be your only clue. Use lowest safe power settings to reduce your airplane noise, but solid flying speed is paramount.

Caribou and Wolf Survey

Surveys of caribou and wolves can usually be lumped together since the wolves will be following the caribou. Additionally, the grizzly/brown bear will prey on new calves as well as injured caribou. (Contrary to popular stories, wolves will take on even the biggest and healthiest of hooved prey such as bull moose in their prime.) The biologist will have ideas on where to start, but once you get "on-track," it's up to you to stay there until you catch up with the animals being tracked. Normally you can't get on-track very well unless there is fairly new snow on the ground.

Tracking wolves is not the easiest thing to do, especially when the snow is hard-packed and wind-swept. If you can establish their probable distant navigational reference point, it could help. However, many times they will simply be going from one hunting valley to the next, or just following a creek. If there are caribou tracks nearby, you can use the 'boo tracks as a back-up when losing the wolf tracks temporarily. Odds are, the wolves will

be near their "dinners on the hoof." They will likely be downwind of the herd. On a sunny day, if you fly on the down-sun side of the tracks, they'll be easier to see with the shadows being cast into the impressions.

Although traveling caribou tend to spread out in ragged short columns, wolves will single-file, each stepping into the leader's foot prints in the snow. The wolf pack will spread into individual tracks when making a line-abreast sweep through brush lines for small prey. This presents the opportunity for you to count the wolves participating in the hunt.

Enforcement Patrol

Unfortunately people violate safeguards established to protect the futures of fish and wildlife. You can expect that during salmon stream survey flights, you will be asked to assist the biologist in looking for violations such as boats fishing in closed waters or during non-fishing hours. In connection with herd animals you'll be looking for evidence of wasted meat or uncontrolled herd shootings. In these cases, scavenger birds such as eagles, magpies, and ravens will congregate at the scene and lead you in when they take to the air upon your noisy approach.

Wolves are shot from the air by poachers when the fur is prime in winter. The poachers will wait for a fresh snow to reveal recent wolf pack activity. First, they will drive the pack into open country

(so they can land), then fly low over each wolf. When an airplane flies low over a wolf, the wolf will ordinarily stop, squat while looking up, and then run off in a different direction. The poacher shoots the wolves during their vulnerable stops. Then, he lands, skins the wolf, takes the pelt and goes on to the next one. Wolf poachers will use small planes on skis or very large tires, as well as higher-powered, larger airplanes on skis and wheel-skis. Your job on enforcement patrol is to carry the enforcement officer in order to discourage such activity as well as to collect evidence and, if you get really lucky, intercept poachers.

It will be helpful to your enforcement officer crew member if you bring your own camera (with 135 mm zoom or bigger) and a notebook for your own notes. Know the passes nearby that lead to airports or towns. This knowledge will stand you in good stead for intercepting fleeing poachers. Dress warmly. Even with a fully insulated airplane, F. Atlee Dodge's superb cabin heater mod, and arctic clothing, including -70 degrees F. "bunny boots," those four-to-six hour patrols get hugely cold.

Poachers will land in some very sporty places, so you must use restraint and be picky about where you'll land to collect evidence. If you can have a GPS receiver along, it could prove valuable to the effort just to know where the evidence is. (GPS and a good camera can also be important in documenting commercial fishing violations.) Learn the location of, and practice using, nearby little-known passes. They could prove to be very handy.

Out of many patrols, we intercepted an airborne poacher only once. We knew the trail was hot because the wolf carcasses were not yet frozen solid and the last wolf he shot was loaded aboard his airplane without skinning. He obviously saw or heard us coming, but we didn't see him in the light snow showers. We headed for a little-known pass that gave us a shortcut to the far side of a more commonly used pass. Visibility was about two miles in light snow. Then we spotted the rascal, closed him from behind, got his number, and then pulled up on his wing for photos. He saw us; I took off a mitten, gave him a universal sign of disapproval and turned away.

A Guiding Principle For Conducting Aerial Wildlife Survey

Wildlife surveys can be hazardous. There is no aspect of wildlife surveys that justifies jeopardizing the safety of flight crews. You can minimize inherent risks by thorough preflight and in-flight planning.

CHAPTER NINE

SURVIVAL FLYING IN EXTREME CONDITIONS

Introduction

This chapter is included in the book because some day you might use this information to deal with unexpected extreme conditions. Any wilderness, by definition, is going to have a paucity of weather reports, upon which forecasts are partly based. That's why, as a wilderness-flying pilot you should take every opportunity to make a PIREP. That PIREP will, in turn, improve the forecasts you receive.

This information, based on actual experiences of the author, might give you ideas when faced with similar situations. First, an important caveat: the applicability of this information to a situation that you might face depends on your type of aircraft as well as the specific conditions. There could be important, disqualifying differences.

I learned from first-hand experience that there are many situations that the wilderness pilot could be facing as a result of rapid, surprising changes in the weather. Those situations can pose big challenges, but they could be manageable when you are armed with information based on the experiences of others.

Nevertheless, the techniques presented here might not work for you and your model of airplane. There is always the possibility that your situation is not the same even though you perceive it to be so. In fact, there is no guarantee that any of the techniques described in this book will always work.

Do everything possible to avoid getting "snookered" by the weather, but don't despair if you do get trapped. Proceed with deliberate caution. Analyze the situation carefully and methodically. Use the outcome-based analysis explained earlier in this book. Then, do what must be done to minimize possible damage to your trusty airplane and its contents. The bush pilot must be creative while remaining practical and steadfast in dire situations.

This chapter also includes information on the effects of extreme cold on engine carburetion and on the airplane structure, as well as places where you can expect to find localized extreme cold in particular terrain.

Extremely High Winds

During his initial years of floatplane and wheel plane flying in the Kodiak/upper Alaska Peninsula and wheels, wheel-skis, and ski flying in the interior Alaska/ Alaska Range areas, the author experienced numerous episodes of dealing with high winds as an air taxi/contract pilot. Those winds had all been up to about thirty-five knots and they were a challenge. But we had solid methods for dealing

with them. Furthermore, flying well over twelve hundred hours a year, we were current -- we were "in tune" with the airplanes we flew. With an experienced passenger who was insistent and willing to endure the rigors of rough air we would usually "cave-in" and go flying. I thought I'd seen it all. I was wrong.

It was the weather in the Lower Alaska Peninsula and Eastern Aleutians that proved me wrong. That area is the birthplace of many unexpected, vigorous low pressure systems. It remains a challenge even to computerized weather forecasting. As a result, that area produces occasional surprise, rapidly developing high winds as well as low ceilings and precipitation of all forms. Being based, for the most part, at the Alaska Department of Fish and Game office near the runway at Cold Bay, from April/May until September/October every year, I had a high workload job to do. So I tried to expand my operational "envelope" as much as possible in order to increase available flight hours.

Consequently; I learned to deal with winds in the thirty-five to forty knot range, even though those speeds were frequently above the stall speeds of the Super Cubs and amphibious Beaver I flew. Because of the special ground handling circumstances I could defy the old dictum that if wind speeds are above stall speed you can't land successfully. Although we normally never planned to go flying when the winds were forecast to be above 30 knots, our peculiarly fortunate ground parking situation allowed us to get around the old

dictum if higher winds surprised us. During 1986 and the following seven years I learned to use the directional characteristics of Cold Bay's high winds. That, combined with the alignment of our wonderful World War II, thirty-foot deep dirt revetment, made for operational flexibility. The revetment is mid-field next to, aligned about thirty degrees to Cold Bay's wide, two-mile-long runway 16-34. Except at the entrance ramp, it has an additional twelve to fifteen feet high dirt rim. (See figure 9-1.)

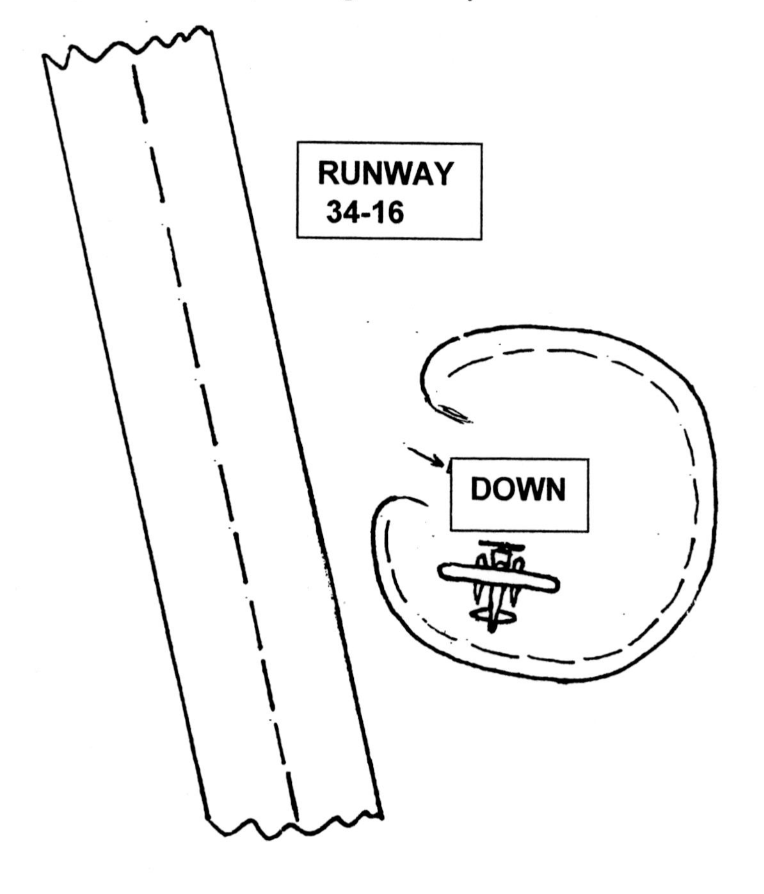

COLD BAY REVETMENT AND RUNWAY
FIGURE 9-1

When the winds are strong at Cold Bay, they usually blow from the directions of 160 degrees magnetic, plus or minus twenty degrees, or 340 degrees magnetic, plus or minus twenty degrees. With the wind from the southeast, I could make my landing approach to runway 16, lining up along the right side. Then while yet high enough, take a little jog left, lining-up with the revetment entrance.

In a Super Cub, I'd reduce power, touch down tail high, well up on the wheels, and raise flaps. (Being on 29" tires made the nose down attitude possible without the big prop striking the runway.) Holding a little power, I'd keep the tail up high until I started feeling the turbulence of the wind being deflected up over the revetment. Then, it was back to idle power, and down the ramp with stick full back and alternating maximum brakes. The Cub was still being buffeted but by then the wind would be much lighter and less organized. That made safe ground handling a feasible proposition.

The southeast wind procedure with the amphibious Beaver was basically the same, except that immediately on touchdown, I would raise flaps and smoothly pitch to put weight on all four wheels. Then I'd ease the control yoke completely forward to ensure constant forward wheel pressure on the surface, staying at minimum angle of attack, as with the Cub.

We avoided using the Beaver with strong northwest winds that required the special procedures described in the following paragraph.

The amphibious Beaver is not directionally stable and rudder-responsive enough for what had to be done with precision. Had it been on "plain" wheels, it would have been worth a try.

With the Super Cub, in northwesterly strong winds, I'd fly up the right side of runway 34, make a slow S-turn right, then left, lining up on the revetment entrance going away. After touchdown high on the wheels, it was flaps up, hold the tail level and ease off smoothly on the power to allow the wind to push us back towards the revetment entrance. I'd use power in lieu of brakes to keep the process slow, while being pushed by the wind. One had to take care to be easy with the rudder, to use ailerons to keep the wings level, and to keep the stick forward until down out of heavy wind effects.

The first time I did this was not planned. In fact, I had continued to be reluctant about northwest winds due to concern about directional control and ability to see enough when the plane was going backwards. But a surprise high wind provided a chance to try it. It turned out to be easier than I had expected. The keys were working hard on precise line-up and landing as short as possible in the revetment entrance ramp. I had to skim by the Beaver, which was parked just south of the lower end of the ramp.

Being able to operate with the conditions described allowed us to do a lot of solo cargo and stream-marking flights in high winds, saving the decent weather for our airborne salmon stream

surveys with biologists. In fact, there were some salmon stream beaches that were simply too short or too steep to land on and erect regulatory markers unless the wind was "snorting" on-shore. It was great fun when going into some of the smaller bays confined by steep terrain. Then, I would have to look out for water spouts (small off-shore tornadoes). Operating in these high wind conditions also emphasized some important principles one should observe:

Don't even think of combining high winds with visibility below about three miles or with ceiling below about 1000 feet. A low ceiling or low visibility will severely limit your choices while navigating. You'll have a hard time keeping a "back door open." And you'll have difficulty avoiding powerful, destructive down drafts. This is very serious and a "big-time" limitation. For example, a Cessna Caravan, piloted by an experienced pilot, was destroyed on the downwind side of a small islet (a hill poking up out of the beach) just west of Cold Bay in a high winds, low ceiling situation.

Wear a helmet and sit low; keep your seat belt snug. In the tough spots, I'd cinch-up the seat belt, sit low, set the power for a speed just above Vy, then use the throttle hand to help brace myself. Wearing a helmet prevented some nasty bangs on the "noggin."

Fly at between Vy and Va, but closer to Vy than Va. Rough air can cause big excursions in your airspeed and put you above Va before you know it.

Be smooth on the controls; positive and firm yet not radical, so you don't over-stress the tail feathers with sudden, heavy loads. (Being below Va doesn't necessarily protect your tail feathers if you jerk the flight controls when you're in hard, choppy air.) Use less than full flaps for landing; carry extra approach airspeed for gusts.

If your landing area is surrounded by trees, expect (and be ready) for the large wind shear robbing you of airspeed as you descend below tree-top level. Prior to landing, analyze the expected wind effects of terrain irregularities that are upwind from your proposed touchdown. There are several places where I've landed that require a downwind landing because of high cross-wind turbulence in the critical touchdown zone.

Wind effects can be severe enough to disqualify an area. If you decide to land, take care; high winds tend to greatly shallow your glide slope and make you low. The remedy is to start with a higher-than-normal glide slope and be ready to add power as you slide onto a normal glide slope.

During climb-out after takeoff, avoid getting nose-high just to slavishly stay on Vx or Vy. Use a lower nose attitude and add an appropriate gust factor to your normal performance airspeed. Then, you are less likely to be robbed of airspeed by unfavorable shears. Consider that a strong head wind will shorten your takeoff run and steepen your climb, so you may not even need to use Vx.

A Super Cub Landing With Sixty-plus Winds

The following took place on the lower Alaska Peninsula in the late 1980s. We were in a belly-tanked Super Cub, on a salmon stream survey flight out of Cold Bay, with the Area Management Biologist in the rear seat. The weather was looking good and was forecast to remain so until the next day. The opposite happened.

About seven hours out of Cold Bay, we had finished the extensive Meshik River system in the Port Heiden area and were headed for Cold Bay, planning to survey a few more small systems on the way. While cruising, I made it a practice to listen to periodic marine forecasts and country music on the AM radio station in Dillingham, on Bristol Bay. The forecasts were holding out for gentle weather.

As we went by Bear Lake, northeast of Port Moller, I saw signs of the opposite. Gibraltar Rock, on the long northeast shore of the lake, was enshrouded with low fracto-cumulus and the winds were kicking up high spray. Johnson's Lodge at the northwest outlet of the lake was also being hit by the same stuff. At that time I could see our ground speed diminishing greatly.

Making it to Cold Bay was out of the question. The strip at Johnson's that was most into the wind would allow no shelter for taxiing. So, we'd have to try and land at the Port Moller gravel runway, three miles from the active cannery (with a Fish and Game cabin nearby).

The Port Moller strip is about three thousand feet long and about one hundred feet wide, with large gravel, and oriented along 190-010 degrees magnetic. At the north end of the runway, on the east side, is a square-log building, used by the Johnsons from Bear Lake as a storage and fuel facility. The building is about fifty feet long, parallel to, and separated from the runway by a gravel loading/parking mat. The runway parallels and is close to the Bering Sea beach. The Port Moller outer bay is about a mile inland from the strip.

It was readily apparent from the outer bay surface that the wind was in the storm category (above 55 knots) from the southeast. The cannery office reported 60 knots with much higher gusts on its anemometer. We cruised south along the beach, staying up at 500 feet AGL, in order to stay out of blowing sand until we could look the situation over and make a plan.

The plan was easy to come by -- our only choice was to land into the wind, across the runway, toward the storage building, and make it come out close without running into the building itself. I knew, from having tied down in the building's lee before, that there were hefty tie-downs close to the building. I had already gained some valuable experience in high winds, and felt ready to take a giant leap, so to speak. Anyway, we didn't have alternatives.

The beach gave no respite with the sand blasting

going on near the surface. But the building did, and then some. Because of the sand blasting, I decided to make only one approach and would go-around only if flight control deteriorated early.

I envisioned that we would encounter a big bounce-wave, then a large down draft with turbulence pummeling the wings as we went close into the lee of the building. As we entered the down draft, I wanted to be on or very close to the gravel, so that any unbalanced shear would cause us to land on a wheel instead of a wing-tip, just in case aileron control was inadequate to get the wing up in time. The illustration shows how I envisioned the wind flow. (See figure 9-2, below.)

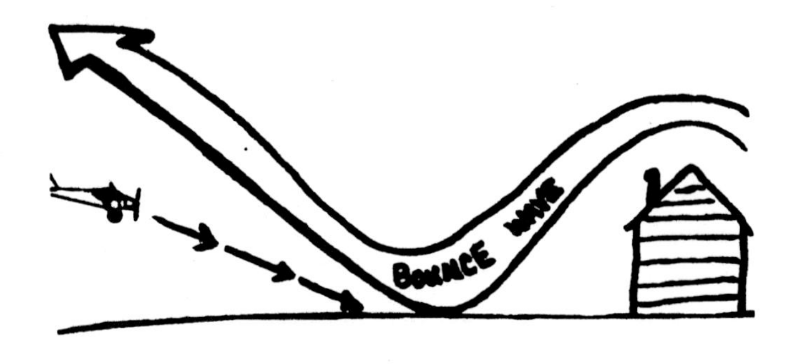

HIGH WIND LANDING
FIGURE 9-2

Throughout the approach it was easy to control the ground speed with throttle. My plan was to hold a ground speed equivalent to that of taxiing slowly. I had only twenty degrees of flaps extended. I wanted to have good rudder control and not too much aerodynamic drag. We felt the bounce-wave much earlier than I had expected. I came far back

on the throttle and dropped the nose quite a bit. I remember saying to the biologist on the intercom, "Brace yourself, Arnie." (Of course, he had been braced from the time I told him what we trying to do.)

Then we got lucky. Just as I felt us come out of the bounce-wave and was adding some power and getting the nose up near level, we entered the down draft. According to plan, I raised the nose more and chopped the power but we got our next thrill: a big left-roll moment from a gust. I countered with right aileron and rudder. My control inputs started to take effect just as we hit on the left main tire.

It was fortunate that the tire pressure was at only six psi (for making convenience stops on soft beaches). Because of the soft tire, our hit on the gravel didn't add much rebound to the right-controls input. Even so, we did a couple of almost uncontrolled lateral bounces back and forth. Then, I slowed the airplane with alternate braking and fed in back stick. Getting stopped with about twenty feet to spare, the wings were rocking violently back and forth, presumably from gusts coming around the ends of the building.

I had to break one of my rules. I asked Arnie to get out of the airplane with the engine running. I told him to stay behind the main lift strut and grab it where it enters the lower surface of the wing. Then he was to try and hold it steady as I taxied closer to the building. It helped. We secured the plane and

endured three miles of stinging sand, dirt, rain, and salty spray blasting us on the four-wheeler that took us to the cannery cabin.

My mistake in that episode was expecting the bounce-wave and down draft to be much closer to the building than they were. Had I better accounted for the strength of the wind, we would have been better prepared for the bounce-wave, down draft sequence. But luck was with us and no damage was done.

Had we been in the amphibious Beaver, I would have made a water landing in a protected cove in nearby Herendeen Bay. The water landing would have been straight in, to a short, shallow, and firm beach. There, I had previously put the wheels down and taxied up to alder brush for shelter.

Emergency Slowing of a Float Plane

Here's an incident that I'm not real proud of but it provided a lot of laughs for my pilot friends. It also provided the author with a good lesson.

It was a rare day in Kodiak with clear skies and light, northwesterly winds. I was fairly new to the U-206 "straight floats" airplane, but that was no excuse for what I did. My immediate chore was to takeoff with full fuel, land on the downtown channel and pick up a load of outbound hunters. As I taxied for warm-up and takeoff on Lilly Lake, I noticed the light surface wind shifting direction frequently. This was common due to the hills just northwest of the

lake. But what I had failed to notice was the wind shifting to a tailwind just as I turned for takeoff. Lilly Lake is short and intolerant of poor technique. Down wind takeoffs qualify as very poor technique. Blithely, like a novice, I turned and "let 'er rip" into my northeasterly takeoff run. And like any eternal optimist, I kept hoping to get up on-the-step but it wasn't happening. I should have aborted the takeoff right then. By the time I gave up the takeoff and chopped power, the remaining lake was very short and rapidly getting shorter. I then put the brain "in gear" and used a technique I'd learned years before.

Applying full nose-down elevator, I rocked the plane forward, digging the bows of the floats into the water. The plane slowed noticeably. As it neared the end of the lake, I went to full up-elevator. The airplane did not lift much but the floats were better positioned for encountering the shallow dirt/gravel slope at lake's end. The plane slid to a quick stop as I shut down the engine.

Chief Pilot Ralph Wright showed up. (Ralph had seen the entire disgusting episode!) Together we placed the plane on a wheeled cradle, pulled it out of the water, inspected it inside and out, and put it back in the water. There were no leaks or damage. With his usual calm, he asked if I wanted to take the flight. I said, "You bet!" Then, with a twinkle in his eye, he asked, "By the way, Hal, what direction is the wind from?" Sheepishly, I gave the correct answer and he sent me on for the flight.

Sudden Loss of Forward Visibility

If you fly in conditions of low ceiling and low visibility, you should always have a plan of actions available to extract you and your airplane.

For example, consider what you would do if you're flying along a coastline where high steep bluffs meet the ocean. I was doing that sort of considering along that kind of coastline one winter in Kodiak. Our Piper Cherokee Six was full of passengers and baggage on the way from the town of Kodiak town to Karluk village. We were experiencing snow showers all the way and they seemed to be getting thicker. I had tried going via Larsen Bay and down the Karluk River, but was blocked by solid stuff as we approached the bend at Shasta Creek. So, at the moment of my contemplation we were going "outside," just past Rocky Point.

We couldn't fly inside the beach because of the bluffs, so I was denied the usual option of putting the water's edge at the center of any required turn-around. I should have turned around right then but decided to hug the bluffs and, if we were forced by weather to turn around, I'd do a one hundred twenty degrees turn to the right, away from the bluffs. Starting from a large crab to stay away from the bluffs, due to the strong onshore wind, I knew that a turn-around of more than one hundred twenty degrees might be too much.

Soon, we had to turn because of a heavy snow

shower. I made a standard rate turn to the right, away from the bluffs. After the turn, the bluffs were not in sight. In fact, nothing was in sight. So, I turned an extra twenty degrees away from the shoreline. (Best give Rocky Point a wider berth.) After thirty seconds I started a descent. Shortly thereafter, we broke into the clear.

Then we went to Larsen Bay Village airstrip for a three-hour wait until the weather improved. The passenger up front was an instrument-rated Private Pilot. He told me he'd never seen that kind of turn-around before. I told him I hadn't either, and then I explained its advantages for the high bluff situation.

There were three big lessons re-learned from that experience: First, it's hard to estimate how far the edge of a snow shower is; so don't push your luck. Second, think in advance about how you will do any required turn-around and tailor it to the wind as well as to the particular terrain you're following. Third, don't wait so long to turn around and you'll avoid the problem in the first place.

A Cheap Lesson

One evening in late May, I was at the Port Moller cannery chow hall, having a 2100 late snack at what they call "mug-up." We were temporarily based at the nearby gravel strip during the impending commercial herring roe season. Suddenly, a young pilot whom I knew to be a fish spotter pilot came in, grabbed a cup of coffee, and sat down across the table. (We'll call him "Bill" but that's not his real

name.) Bill was gulping coffee and chain-smoking cigarettes. Obviously, he had a tale to tell, so I asked him, "what's up?"

Bill had just completed a very sporting flight. He had been flying low along the Bering Sea beach trying to get to Port Moller before the impending harvest season kicked off for that relatively small fishery. He hadn't done very well in the big Bristol Bay fishery and was desperate to make a few bucks at Port Moller. Thus, he was primed to push his luck as the fog started to come ashore southwest of Port Heiden and Ilnik. Soon, he was down "on-the-deck" starting to consume some of those nine lives that every fish spotter pilot must start with.

Then, suddenly he saw the beach terminate close ahead with a solid bluff going up into the thick fog layer. He racked his Super Cub to the right to miss the bluff, pulled up to miss a big black rock coming out of the fog, then pushed nose down and turned left towards surf that he saw leading to a continuation of the beach. He pulled out a tad late, landed inadvertently, bounced, and chopped his power for a bumpy roll-out, just missing a few huge walruses on the beach. Then he waited until the fog lifted. It was obvious that he was being introduced to Cape Seniavin, late spring hangout for bull walruses. What a great learning experience for a young pilot without his paying the ultimate price!

We talked about the lessons to be learned and various methods of measuring visibility. (I like the method of timing from first sighting of a ground

object until you reach it at a steady speed.) We also discussed different ways to plan your turn-around in sudden loss of visibility or maneuvering room. I introduced him to my 1:250,000 Coast and Geodetic Survey charts and we looked over his route on them. Then we drove out to the strip on a three-wheeler and checked his airplane for damage. He had gotten by without penalty.

The Climb Escape Option

Sudden loss of flight visibility due to cloud/fog can usually be handled safely by a quick reversal of course. But what if you're in a pass that's too narrow in which to turn around -- a one-shot pass? (As mentioned in an earlier chapter, an ordinary pass can be turned into a "one-shot" by high tail winds.) If (and this is a big "if") the terrain ahead and along your projected course ahead permits you to do so, a last-ditch maneuver to save yourself could be a straight-ahead climb through IMC to safe altitude.

Generally, you'd want this to be up to known on-top VMC and in uncontrolled airspace. To repeat, you need the certainty of clearing any terrain that is ahead of you! In ten thousand hours of flying the hills of Alaska, the author has had to use this maneuver only twice -- both of which meant that I had failed to fully appreciate what I was getting into, otherwise they wouldn't have been necessary.

The first time was when I was hauling a friend in a Cessna 180 on wheel-skis from Kantishna to

Fairbanks. On the way out for the pick-up, I had noticed that the Toklat drainage had good weather but Bearpaw River and westward was socked in with low clouds and fog. From Kantishna, I planned to go up the North Fork of Moose Creek, then up Willow Creek over a saddle and down Myrtle Creek to the Clearwater Fork of Toklat River. Then it would be almost a straight shot to Fairbanks. (You need a chart of the area north of Mt. McKinley to fully appreciate what was going on here.)

After we departed, Myrtle Creek was socked in, so I stuck with the North Fork a couple miles to the east and headed for a little one-shot V-pass leading straight into Clearwater Fork.

There was a heavy layer of snow that made the hills beautiful but it also helped create the sudden "white-out" after we entered the pass. I had already planned a straight-ahead escape climb if needed, and so it was a simple matter to take a climb pitch attitude on instruments and shove the throttle full forward. To have attempted a turn-around would have been foolish -- there were no visual references, and not enough room.

My friend asked, "Is this intentional?" I knew he meant "Was this planned?" but, embarrassed, I answered him literally with, "Yes, it is." We broke out with Kankone Peak poking above the tops five miles to the left behind our wing and Mt. Sheldon at two o'clock, ten miles. In this case, I hadn't known

for sure that the one-shot pass was clear of weather all the way through. Therefore, I should have rejected the pass, gone back to the Toklat and followed it northeastward. Good lesson, re-learned.

The other time I used the climb escape maneuver was during an empty return with a Piper Saratoga wheel plane from Old Harbor to Kodiak. State Airport was VFR but reporting clouds southeast.

As I crossed Ugak Bay on the east side of Kodiak Island, the radio came alive with traffic calling the tower from Kalsin Bay. They were headed across the larger Chiniak Bay to the municipal airport -- just the route I was anticipating once I crossed from Ugak over Lake Miam and Summit Lake then downhill into Kalsin Bay. I had flown this route many times and knew that once I got to Summit Lake, if I could see a chunky little Sitka Spruce tree on a knob to the northeast, ceiling and visibility were good enough to make it safely to Kalsin Bay.

As I reached Summit Lake and looked northeast, the little tree appeared. Then, while I was looking at it, the tree disappeared, enveloped in a fog layer. I did a one-eighty to the left in the wide valley, looking back to see clouds on-the-deck obscuring Ugak Bay. It was time to climb. I didn't want to get close to the controlled airspace at Kodiak, so I climbed southerly. I broke out on top near Pasagshak Bay and swung wide to the northeast. Then it was a simple matter of clearing the upper shelf of clouds a little south of Cape Chiniak and going on to municipal airport in VMC.

Cold Weather Effects

Most pilots who fly in the average North American winter understand all the things one should do to keep winter flying the joy that it is. When the author first started flying the interior Alaskan winter, he read all the material he could lay hands on concerning operations at below-zero Fahrenheit. The reading was interesting and informative, but it could have been more detailed for my purposes.

But I did learn, for example, that the aluminum alloy used to make airplanes strong yet light weight starts getting weak below minus 30 degrees F. By the time it reaches minus 40 degrees, it is not only weakened substantially, but also has become brittle. Therefore, I established minus 30 degrees as our minimum operating temperature. I learned from experience and paying attention to the OAT gage, that one can takeoff when it's minus 30 at the home strip, start a survey flight and easily go into low valleys where the temp would be minus 55 degrees. That's going into harm's way, with respect to airframe <u>and</u> engine.

For example, a friend once asked for my help in shuttling loads of cargo from Fairbanks to a cabin inside the Alaska Range. The temperature at our strip was minus forty-four, my neighbor at a lower elevation was looking at minus sixty-two on his gage, and it was minus forty-three in Fairbanks. I just could not bring myself to say "okay." He decided to try it anyway. He taped off the engine air

inlets, started flying and ended up landing his ski plane on Chena River ice with a dead engine plugged with frozen oil. We finished the job in kinder weather just two days later.

A Quick Cold Soak of the Engine

The material I studied advised against doing touch-and-goes when the OAT is below plus 20 degrees F. That's understandable because of the rapid engine cooling that can take place on a low-power approach. But I reaffirmed, the hard way, that it's a good idea to get the oil temp into the operating range before you takeoff in temperatures below zero. I had flown a Super Cub on replacement skis to a winter lodge strip on the Wood River (the one in the Alaskan interior that runs into the Tanana River east of Nenana). The Wood River at that inner-Range point is more like a creek. The temperature was minus 25 degrees. It was a brief stop, so I didn't throw the usual engine blanket over the cowl.

On takeoff, just after my skis cleared brush at the upwind end, the engine sputtered and died. I did a jog left and landed on the snow-covered ice of the river. After assuring that there was nothing unusual in the engine compartment, I re-started and warmed up the engine, shut down, threw the blanket on and let it get a good heat-soak of about five minutes. After that, the engine operated normally, oil temperature came up quickly and we departed.

Cold Weather Carburetion

You may already know that, in frigid weather, some float carburetor-equipped engines will tend to lean-out at the high-cruise throttle setting. I've never seen reading material on this, but it doesn't take a rocket scientist to figure out what's going on. It is due to the very dense air created by very low temperatures. The carburetor main fuel jet is not able to supply enough gas to match the dense air coming through the carburetor. And so, the mixture becomes too lean at and above high cruise throttle setting.

With some (not all) carburetors, as the throttle is pushed beyond cruise setting, an economizer jet opens, allowing what normally is extra "cooling fuel" in the near-takeoff and takeoff throttle positions. This extra fuel is needed to prevent high cylinder temperatures that invite pre-ignition and detonation. But at arctic temperatures, this extra fuel combined with the dense air can be instrumental in over-boosting some of the higher compression engines. In this situation, you should monitor the manifold pressure and observe its limits, especially when using higher power settings.

In temperatures below about minus 10 degrees, if you push the throttle forward slowly on takeoff and the engine runs a little rough in the high-cruise throttle position, put the carburetor heat on, to the point where the engine runs smooth, or the carburetor heat is full on. If it continues to run a

little rough, reduce the throttle until the engine runs smoothly. To avoid the rough-running transition simply put carb heat "on" before going to full throttle. Then throttle forward until reaching full throttle or the edge of the rough range, whichever occurs first. After setting the throttle, you then ease off on the carb heat until the engine ran its smoothest, if it weren't already purring. This technique should work for all carburetors. In any case, see your engine operating manual.

The important things to know are that, in frigid temps, it's easy to over-boost your bigger engines and possibly some of the so-called smaller engines, especially those that have been upgraded in their horsepower output such as the 160 hp upgrade of the O-320. Further, depending on how effective your carburetor heat is, you do have some control over the "internal density altitude" at which your engine is working. Finally, never tolerate a rough-running engine. There's always a reason for any abnormally operating engine.

Blowing Snow and Flight Controls

As the penultimate subject on cold weather effects, you should be aware that blowing snow can get inside your flight controls, change their mass balance, and give you the ride of your life as you get airborne. I once noticed an unusual heaviness in the controls of a U-206 on wheel-skis as I did a pre-takeoff flight control check. We immediately took it into our heated hangar, suspecting that snow and

had collected in the controls. (We had seen a lot of snow and high winds for three days previous and the 206 had been outside, hooked up to engine and cabin electric heaters.) When the hangar heat started melting all the packed snow in the ailerons, elevators, and flaps, the source of that extra weight was evident.

That airplane would have been a chore to fly with fluttering ailerons. I've heard horror stories of people having them come off the hinges due to snow packed within the fluttering ailerons. If you ever wonder what's wrong with the controls before or after a cold weather takeoff, consider this possibility. If you're airborne and the ailerons look fuzzy, slow down! They're fluttering and getting ready to come unglued. Slow down and make a careful landing approach.

Cold Weather Gyro Spin-Up

Finally, watch out for slow spin-up of your instrument gyros in cold weather. Give your attitude gyro plenty of time to get up to speed, to minimize internal damage to the gyros, and also to give you reliable instrument references. Slow gyro spin-up can be avoided by having an effective but safely positioned electric heater for the cabin. Also, of course, have one for your engine compartment. Consider the following episode.

It was Christmas Eve and the lady on the telephone at McKinley Park Village was doing her best to convince me to take her to join her husband

at their lakeside cabin far to the west. The weather was snow showery but we would give it a try though I would return after dark. After making sure by HF radio that the weather at the cabin was good enough and there was no overflow on the lake in front of their cabin, we departed. The flight out was routine, but I had to stop for fuel at Kantishna on return that evening. I had depleted my Kantishna fuel cache the day before, so I had to ask Dan Ashbrook (gold miner, trapper, pilot, and friend) for a fifteen gallon loan in order to have a full reserve.

Dan, as usual, was a great help with the fuel and then parked his snow machine with lights across the upwind end of the snow-packed strip for my takeoff. The engine of the wheel-ski Cessna180 was still warm due to the engine blanket. But the cabin was frigid. It was turning into a "three-dog night" with the temperature already at minus twenty-five.

My plan was to takeoff to the southeast, into the light breeze, and make a climbing left turn prior to reaching the notch that Moose Creek makes as it heads north into the Minchumina Flats. If I delayed the turn and flew through the notch there might be low clouds lurking in the hemmed-in area beyond. Dan had mentioned some clouds moving in from North Fork. I wasn't about to takeoff downwind on skis, not because of performance reasons, but to keep a late abort as a practical option on takeoff.

The takeoff went normally, until just after liftoff, when I looked at the attitude gyro. It showed the horizon at the <u>bottom</u> of the indicator. I cross-checked the VSI, which showed a normal Vx climb rate. Airspeed was staying on Vx, so I was looking at the symptoms of an under-speed attitude gyro. (I've explained to student pilots many times how a failed or under-speed gyro horizon will drop to the bottom when you rotate for liftoff; inviting you to put the nose down and fly into the terrain.)

As I crossed over Dan's lights, I could see there was adequate altitude for a left turn to northwest. I had a decision to make because I felt leery about relying on the turn coordinator which could also be under-speed. However, in the dark, I could just faintly see the snow-covered hills to the left of the Moose Creek notch. I immediately turned off all of the airplane's lights, could see the hills well, and made the turn by visual reference to the snowy slope with its dark patches of vegetation.

During the remainder of the departure the instruments gradually returned to normal. I was glad about that. Snow squalls blocked my return to our strip south of the McKinley Park entrance. After making an instrument approach into Fairbanks I had to wait for a Christmas Day return to our cabin inside the Alaska Range.

Thereafter, if the cockpit was really cold, I was more patient about giving the gyros a better chance to get ready for flight. Turning all lights off to

improve outside night vision in an emergency was something I learned as a youngster flying Navy fighters. My random-access memory is a bit ancient but it met the challenge of that cold-weather, slow gyro spin-up experience.

Remember, make whatever works safely work for <u>your</u> safety. Good Luck!

The Author